T0214807

Lecture Notes
in Business Information Processing   **388**

Series Editors

Wil van der Aalst ⓘ
  RWTH Aachen University, Aachen, Germany
John Mylopoulos ⓘ
  University of Trento, Trento, Italy
Michael Rosemann ⓘ
  Queensland University of Technology, Brisbane, QLD, Australia
Michael J. Shaw
  University of Illinois, Urbana-Champaign, IL, USA
Clemens Szyperski
  Microsoft Research, Redmond, WA, USA

More information about this series at http://www.springer.com/series/7911

Danielle Costa Morais ·
Liping Fang · Masahide Horita (Eds.)

Group Decision
and Negotiation

A Multidisciplinary Perspective

20th International Conference
on Group Decision and Negotiation, GDN 2020
Toronto, ON, Canada, June 7–11, 2020
Proceedings

 Springer

Editors
Danielle Costa Morais ⓘ
Universidade Federal de Pernambuco
(UFPE)
Recife, Brazil

Liping Fang ⓘ
Ryerson University
Toronto, ON, Canada

Masahide Horita ⓘ
University of Tokyo
Kashiwanoha, Japan

ISSN 1865-1348 ISSN 1865-1356 (electronic)
Lecture Notes in Business Information Processing
ISBN 978-3-030-48640-2 ISBN 978-3-030-48641-9 (eBook)
https://doi.org/10.1007/978-3-030-48641-9

This Springer imprint is published by the registered company Springer Nature Switzerland AG
The registered company address is: Gewerbestrasse 11, 6330 Cham, Switzerland

Preface

The Annual International Conferences on Group Decision and Negotiation have provided a stimulating environment for the dissemination of the state-of-the-art knowledge in the field of group decision and negotiation, allowing for intense discussions among participants and the exchange of ideas and critical comments for further improvement. The series of conferences has been held every year since 2000 (with one exception, in 2011).

The 20th International Conference on Group Decision and Negotiation (GDN 2020) was scheduled to be held at Ryerson University, Toronto, Canada, during June 7–11, 2020. However, due to the COVID-19 pandemic, when we faced a global health crisis, the Council of the Group Decision and Negotiation (GDN) Section, Institute for Operations Research and the Management Sciences (INFORMS), and GDN 2020 general and program chairs most regretfully decided to cancel GDN 2020 on March 25, 2020. The safety and wellbeing of our GDN community members and their families are of paramount importance.

At the time of conference cancellation, the preparation for GDN 2020 almost reached the final stage. We received 74 submissions. Notwithstanding unprecedented circumstances, we remained committed to the publication of conference proceedings.

In total, 74 papers were classified into 7 different streams, covering a wide range of topics related to group decision and negotiation. After a thorough review process, 14 of these papers were selected for inclusion in this volume entitled *Group Decision and Negotiation: A Multidisciplinary Perspective.*

This volume is organized according to four main streams of the conference that demonstrate the variety of research that was submitted to GDN 2020:

- The stream on "Conflict Resolution" brings together different research and application areas in investigating strategic conflict, including the graph model for conflict resolution (GMCR), composition of probabilistic preferences, and analysis of disputed territories
- The "Preference Modeling for Group Decision and Negotiation" stream focuses on methodological issues for supporting groups of decision-makers and negotiators in eliciting preferences
- The "Intelligent Group Decision Making and Consensus Process" stream includes analyses of managing incomplete information, opinion dynamics, and decision rule aggregation approaches
- The stream on "Collaborative Decision Making Processes" contains an ontology for collaborative decision making and a novel decision support system for collaborative project ranking

The preparation of this volume required the efforts and collaboration of many people. In particular, we would like to thank the honorary chair of GDN 2020, Gregory Kersten, and the general chairs of GDN 2020, Keith W. Hipel, Adiel Teixeira de

Almeida, and Rudolf Vetschera, for their contributions to the GDN Section and GDN 2020. Special thanks also go to all the stream organizers: Liping Fang, Keith W. Hipel, and D. Marc Kilgour (Conflict Resolution); Tomasz Wachowicz and Danielle Costa Morais (Preference Modeling for Group Decision and Negotiation); Zhen Zhang, Yucheng Dong, Francisco Chiclana, and Enrique Herrera-Viedma (Intelligent Group Decision Making and Consensus Process); Pascale Zaraté (Collaborative Decision Making Processes); Bilyana Martinovski (Emotion in Group Decision and Negotiation); Haiyan Xu, Shawei He, and Shinan Zhao (Risk Evaluation and Negotiation Strategies); and Mareike Schoop, Philipp Melzer, and Rudolf Vestchera (Negotiation Support Systems and Studies (NS3)).

We are very grateful to the following reviewers for their timely and informative reviews: Bismark Appiah Addae, Luis Dias, Michael Filzmoser, Eduarda Frej, Yuan Gao, Bingfeng Ge, Yu Han, Miłosz Kadziński, Ginger Ke, Sabine Koeszegi, José Leão, Haiming Liang, Yating Liu, Yoshinori Nakagawa, Daniel Nedelescu, Hannu Nurmi, Simone Philpot, Leandro Rego, Annibal Sant'anna, Mareike Schoop, Maisa M. Silva, Takahiro Suzuki, Junjie Wang, Shikui Wu, Hengjie Zhang, Jinshuai Zhao, Shinan Zhao, and Yi Xiao.

We would also like to thank Ralf Gerstner, Alfred Hofmann, Christine Reiss, and Aliaksandr Birukou at Springer for the excellent collaboration.

April 2020   Danielle Costa Morais
                                                                     Liping Fang
                                                                  Masahide Horita

Organization

Honorary Chair

Gregory Kersten   Concordia University, Canada

General Chairs

Keith W. Hipel   University of Waterloo, Canada
Adiel Teixeira de Almeida   Federal University of Pernambuco, Brazil
Rudolf Vetschera   University of Vienna, Austria

Program Chairs

Liping Fang   Ryerson University, Canada
Danielle Costa Morais   Federal University of Pernambuco, Brazil
Masahide Horita   University of Tokyo, Japan

Program Committee

Melvin F. Shakun	New York University, USA
Adiel Teixeira de Almeida	Federal University of Pernambuco, Brazil
Amer Obeidi	University of Waterloo, Canada
Bilyana Martinovski	Stockholm University, Sweden
Bo Yu	Dalhousie University, Canada
Bogumił Kamiński	Warsaw School of Economics, Poland
Danielle Costa Morais	Federal University of Pernambuco, Brazil
Ewa Roszkowska	University of Białystok, Poland
Fran Ackermann	Curtin Business School, Australia
Fuad Aleskerov	National Research University HSE, Russia
Gert-Jan de Vreede	University of South Florida, USA
Ginger Ke	Memorial University of Newfoundland, Canada
Gregory Kersten	Concordia University, Canada
Haiyan Xu	Nanjing University of Aeronautics and Astronautics, China
Hannu Nurmi	University of Turku, Finland
João Clímaco	University of Coimbra, Portugal
John Zeleznikow	Victoria University, Australia
José Maria Moreno-Jiménez	Zaragoza University, Spain
Keith W. Hipel	University of Waterloo, Canada
Kevin Li	University of Windsor, Canada
Liping Fang	Ryerson University, Canada
Love Ekenberg	Stockholm University, Sweden

Luis Dias	University of Coimbra, Portugal
Maisa Mendonça	Federal University of Pernambuco, Brazil
D. Marc Kilgour	Wilfrid Laurier University, Canada
Mareike Schoop	Hohenheim University, Germany
Masahide Horita	University of Tokyo, Japan
Pascale Zarate	Université Toulouse 1 Capitole, France
Przemyslaw Szufel	Warsaw School of Economics, Poland
Raimo Hamalainen	Aalto University, Finland
Rudolf Vetschera	University of Vienna, Austria
Rustam Vahidov	Concordia University, Canada
Sabine Koeszegi	Vienna University of Technology, Austria
ShiKui Wu	Lakehead University, Canada
Tomasz Szapiro	Warsaw School of Economics, Poland
Tomasz Wachowicz	University of Economics in Katowice, Poland
Yufei Yuan	McMaster University, Canada
Jing Ma	Xi'an Jiaotong University, China

Contents

Intelligent Group Decision Making and Consensus Process

Collaborative Decision Making Processes

Conflict Resolution

Nash Stability in a Multi-objective Graph Model with Interval Preference Weights: Application to a US-China Trade Dispute

Jingjing An[1,2(✉)], D. Marc Kilgour[3], Keith W. Hipel[2],
and Dengfeng Li[4]

[1] School of Economics and Management, Fuzhou University,
Fuzhou 350108, Fujian, China
1126560785@qq.com
[2] Department of Systems Design Engineering, University of Waterloo,
Waterloo, ON N2L 3G1, Canada
[3] Department of Mathematics, Wilfrid Laurier University,
Waterloo, ON N2L 3C5, Canada
[4] School of Economics and Management, University of Electronic Science
and Technology of China, Chengdu 611731, Sichuan, China

Abstract. In many real-world conflict situations, decision-makers (DMs) integrate multiple objectives rather than considering just one objective or dimension. A multi-objective graph model (MOGM) is proposed to balance each DM's objectives in both two-DM and multi-DM conflicts. To identify Nash stability in MOGMs, a comprehensive preference matrix with weight parameters on objectives is developed for each DM, along with a unilateral move matrix including preference weights (UMP). Then, considering the subjective uncertainty of DMs, interval numbers are used to represent the degree of uncertainty of preference. Subsequently, Nash equilibria and interval Nash equilibria are developed for MOGMs, and the dependence of these equilibria on weights is shown. To illustrate how MOGM can be applied in practice and provide valuable strategic insights, it is used to investigate a US-China trade dispute model. The stability results suggest potential strategic resolutions of bilateral trade disputes, and how DMs can attain them. The case analysis process suggests that a peaceful settlement of the dispute may be achievable.

Keywords: Multi-objective graph model (MOGM) · Preference weight · Interval preference · Nash equilibrium · US-China trade dispute

1 Introduction

Conflicting interests and objectives are a perpetual concern of economics and other social sciences. There are essentially three kinds of conflict: conflicts among several decision makers (DMs), conflicts within an individual, and conflicts within and among individuals. The first kind of conflict is studied in conventional game theory, where each DM tries to maximize a scalar payoff. The second kind can be seen as the subject of individual decision theory. The third kind, involving several DMs as well as within

© Springer Nature Switzerland AG 2020
D. C. Morais et al. (Eds.): GDN 2020, LNBIP 388, pp. 3–20, 2020.
https://doi.org/10.1007/978-3-030-48641-9_1

each DM, is modeled by a multi-objective game, which reflects that DMs in the game may have several objectives, possibly conflicting.

In the study of multi-objective non-cooperative games (MONCGs), Blackwell [1] first considered vector payoffs and formalized the two-person zero-sum vector matrix game, but addressed only the minimax property. Shapley [2] showed the existence of strategic equilibria assuming each DM will choose a weakly efficient or efficient solution given the choices of the rivals, while Zeleny [3] dealt with the same problem by using linear multi-objective mathematical programming. Borm et al. [4] considered the general two-person bi-matrix game and studied its comparative statics. Charnes et al. [5] proposed the more general n-person MONCG, but limited all DMs' choices to a cross-constrained set. Zhao [6] defined cooperative, non-cooperative, hybrid and quasi-hybrid solution concepts for multi-objective games and proved their existence. Yu [7] also studied the existence of Nash equilibrium and Pareto equilibrium for MONCG. Most of these studies define and prove existence for various solutions, but there are often no effective computational techniques to obtain the equilibrium solution.

Imprecision or fuzziness is inherent in human judgment, and some literature on MONCGs under uncertainty incorporates the concepts of fuzzy set, grey number and probability. Assuming that a DM has a fuzzy goal for each objective which can also be interpreted as a DM's degree of satisfaction for a payoff, Nishizaki and Sakawa [8] studied multi-objective fuzzy two-person zero-sum and non-zero-sum games. They considered the relation of equilibrium solutions for multi-objective two-person games combining fuzzy goals with the Pareto optimal equilibrium solutions defined in Borm [4]. However, in existing research on MONCGs, the payoff in each state is interpreted as a utility value. When game theory is applied to real world problems, it is often difficult to assess utilities exactly, but it is easier to determine the relative payoffs, or order of preference, of the states.

The graph model for conflict resolution (GMCR) is a flexible and comprehensive methodology for systematically investigating strategic conflicts, in which multiple DMs dynamically interact with each other in terms of potential moves and counter moves, in order to fare as well as possible [9–11]. Xu et al. [12, 13] devised matrix representations for calculating individual stability and equilibria for GMCR. Explicit algebraic formulations allow users to develop algorithms conveniently in order to assess the stabilities of states and permitted new solution concepts to be integrated into the decision support system GMCR II [14, 15]. Li et al. [16] proposed a new preference structure for the graph model to handle uncertainty in DMs' preferences and redefined several solution concepts with preference uncertainty. Bashar et al. [17] and Hipel et al. [18] developed a methodology to model and analyze a conflict with fuzzy preferences. Ke [19, 20] designed a multiple criteria decision analysis approach and incorporated an analytic hierarchy process to capture the relative preference information of a DM involved in a conflict through defining fuzzy preference relation. The ideas of grey and probabilistic preferences were also incorporated into the graph model methodology from different viewpoints by Kuang et al. [21] and Rego and dos Santos [22]. However, MONCG with fuzzy preference has not been studied until now.

In this paper, a multi-objective graph model (MOGM) is proposed to balance each DM's objectives in both two-DM and multi-DM games. Two types of Nash equilibria are developed for MOGMs; the dependence of equilibria on weights of objectives is shown. Our paper differs from the literature in the following three aspects.

We mainly focus on investigating the MONCG. This is the first use of GMCR to study this kind of conflict model. A multi-objective graph model (MOGM) is proposed and the Nash equilibrium solution method is given.

Compared with the existing matrix representation of preference, which requires three binary relations, the proposed matrix representation of crisp preference is more intuitive and requires only three values. We use the values 1, 0, and −1 to represent preference by a DM: positive preference, indifference, and negative preference.

Considering the subjective uncertainty of DMs, interval numbers are used to measure the degree of uncertainty of preference. In other words, each cell in the preference matrix is made up of interval numbers, which convey the uncertainty, subjectivity, and linguistic nature of DMs' judgments.

The US-China trade dispute is analyzed from the perspective of the graph model. Three scenarios are used to describe for the bilateral trade dispute and evaluated in terms of the preference of the two sides. In each scenario, a comprehensive analysis considering both short-term and long-term objectives is conducted.

The remainder of the paper is as follows. In Sect. 2, the basic structure of a graph model is reviewed and the multi-objective graph model is proposed. A comprehensive preference matrix with weight parameters on objectives is developed for each DM, along with a unilateral move matrix including preference weights (UMP). Furthermore, the Nash equilibrium solution method of MONCG, both two-DM and multi-DM, is given. In Sect. 3, the US-China trade dispute is introduced briefly and the MOGM methodology is used to analyze it. In Sect. 4, MOGM with interval preference weights is established to determine the interval Nash equilibrium solution. In Sect. 5, the models and methods of this paper are illustrated with the US-China trade dispute. Finally, some conclusions and ideas for future work are provided in Sect. 6.

2 Graph Model

2.1 Graph Model with Simple Preference

The key ingredients in a classical graph model are the DMs, states or scenarios that could take place, and the preferences of each DM [9, 10]. These ingredients are explained in detail followed by the definitions of reachable lists for a DM. Moreover, Nash stability is formally defined, which determines whether a state is stable for a DM.

A n-DM graph model is a structure $G = \; <N, S, \{A_t, \succeq_t, t \in N\} >$, where

(1) $N = \{1, 2, \cdots, n\}$ is the set of DMs.
(2) $S = \{s_1, s_2, \cdots, s_m\}$ is a nonempty, finite set, called the set of feasible states or situations.
(3) For each DM t, $(s_i, s_j) \in A_t$, $i, j = 1, 2, \cdots, m$ means that DM t can move from state s_i to s_j in one step, where A_t is DM t's set of all oriented arcs.

(4) For each DM t, $\succeq_t$ is a relation on S that indicates the preference between states of DM t. $\succeq_t$ is assumed to be irreflexive, transitive and complete. $s_i \succeq_t s_j$ means that DM t be indifferent or prefer state s_i than state s_j.

For $t \in N$, DM t's unilateral moves (UMs) and unilateral improvements (UIs) are sets $R_t(s_i) = \{s_j \in S : (s_i, s_j) \in A_t\}$ and $R_t^+(s_i) = \{s_j \in S : (s_i, s_j) \in A_t, s_j \succ_t s_i\}$. For matrix representation of UMs, DM t's UM matrix is an $m \times m$ matrix, $\mathbf{J}_t$, with (s_i, s_j) entries

$$\mathbf{J}_t(s_i, s_j) = \begin{cases} 1, & if (s_i, s_j) \in A_t, \\ 0, & otherwise. \end{cases}$$

Note that $\mathbf{J}_t(s_i, s_j) = 1$ if and only if DM t can move from state s_i to s_j in one step. In other words, $(s_i, s_j) \in A_t$ [13].

Several preference relation matrices $\mathbf{P}_t^+$, $\mathbf{P}_t^-$, and $\mathbf{P}_t^=$ are defined as

$$\mathbf{P}_t^+(s_i, s_j) = \begin{cases} 1, & if s_j \succ_t s_i, \\ 0, & otherwise, \end{cases}$$

$$\mathbf{P}_t^-(s_i, s_j) = \begin{cases} 1, & if s_j \prec_t s_i, \\ 0, & otherwise, \end{cases}$$

and

$$\mathbf{P}_t^=(s_i, s_j) = \begin{cases} 1, & if s_j \sim_t s_i, \\ 0, & otherwise. \end{cases}$$

where $\mathbf{P}_t^+(s_i, s_j) = 1$ in the preference matrix indicates that DM t prefers state s_j to state s_i, while zero entry $\mathbf{P}_t^+(s_i, s_j) = 0$ indicates that DM t either prefers s_i to s_j or is indifferent between s_i and s_j. $\mathbf{P}_t^-(s_i, s_j)$ and $\mathbf{P}_t^=(s_i, s_j)$ can be interpreted similarly [13].

In contrast to Xu's matrix approach to preference [13], which requires three binary relations, we use only three values, 1, 0, and -1, to express a more intuitive preference matrix. We will use this matrix to represent Nash stability in n-DM graph model.

Definition 1. For a graph model G, the preference matrix for DM t is an $m \times m$ matrix, $\mathbf{P}_t$ with entries

$$\mathbf{P}_t(s_i, s_j) = \begin{cases} 1, & if s_j \succ_t s_i, \\ 0, & if s_j \sim_t s_i, \\ -1, & if s_j \prec_t s_i. \end{cases}$$

In the preference matrix, $\mathbf{P}_t(s_i, s_j) = 1$ indicates that DM t prefers state s_j to state s_i, $\mathbf{P}_t(s_i, s_j) = 0$ indicates that DM t is indifferent between s_i and s_j, while $\mathbf{P}_t(s_i, s_j) = -1$ implies that DM t prefers state s_i to state s_j.

For a graph model G, the UM matrix including preference information (UMP) for DM t can be calculated by

$$\mathbf{H}_t = \mathbf{J}_t \circ \mathbf{P}_t, \tag{1}$$

where "$\circ$" denotes the Hadamard product. Note that $\mathbf{H}_t$ is an $m \times m$ matrix. The (i,j) entry in the matrix $\mathbf{H}_t$ is $\mathbf{H}_t(i,j) = \mathbf{J}_t(i,j) \circ \mathbf{P}_t(i,j)$.

The logical definition of Nash stability of the graph model for conflict resolution is given as follows.

Definition 2 [13]. Let $t \in N$ and $s_i \in S$. s_i is Nash stable for DM t iff $R_t^+(s_i) = \emptyset$.

Theorem 1. State $s_i \in S$ is Nash stable for DM t iff $\mathbf{H}_t(i,j) \leq 0$ for all $j \neq i, j = 1, 2, \cdots, m$.

Proof: If state $s_i \in S$ is Nash stable for DM t, then according to Definition 2, there is no UI for DM t to any other state. From Eq. (1), we derive that each value in the i th row of the matrix $\mathbf{H}_t$ is less than or equal to zero. And vice versa. □

If each DM finds that he or she cannot do better than to stay in the current state, it is a Nash equilibrium [23]. According to Theorem 1, if the state $s_i \in S$ is Nash stable for all DMs, $R_t^+(s_i) = \emptyset$ for all $t \in N$. This indicates that starting in state s_i no DM will change their current strategy, making stable for all DMs.

Note that if a state $s_i \in S$ is Nash stable for all DMs $t \in N$, then s_i is a Nash equilibrium.

2.2   Graph Model with Multiple Objectives

In many real conflict problems, there are not only conflicts among DMs but also multiple conflicting objectives within individuals. For example, the orders of preference of states may be different according to different objectives. Alternatively, each DM may be a "team". This is a multi-objective game, which happens whenever DMs in a game have multiple objectives. The structure of the multi-objective graph model (MOGM) requires a graph model satisfying conditions (1), (2), and (3), with (4), replaced by a multi-objective structure in which $O = \{o_1, o_2, \cdots, o_K\}$ is the set of objectives that all DMs might choose in the n-DM graph model.

Assume that the weight on the objectives of DM t is $\omega_t = (\omega_{t1}, \omega_{t2}, \cdots, \omega_{tK})$. In particular, if DM t does not have objective o_k, then weight $\omega_{tk} = 0$. The structure of the n-DM graph model with multiple objectives, which is denoted by $MOGM = <N, S, O, \{A_t, t \in N\}, \{\succeq_t^{o_k}, t \in N, o_k \in O\}>$, where $\succeq_t^{o_k}$ is a preference for DM t reflecting objective o_k.

The preference matrix of DM t on objective $o_k, k = 1, 2, \cdots, K$ is denoted by an $m \times m$ matrix $\mathbf{P}_{tk}$. In order to incorporate all objectives, the comprehensive preference matrix $\mathbf{P}_t$ with parameter ω_t of DM t is determined by

$$\mathbf{P}_t(\omega_t) = \sum_{k=1}^{K} \omega_{tk} \mathbf{P}_{tk}. \tag{2}$$

Then, the UM matrix including preference weights of DM t can be calculated by

$$\mathbf{H}_t(\omega_t) = \mathbf{J}_t \circ \mathbf{P}_t(\omega_t), \qquad (3)$$

where "$\circ$" denotes the Hadamard product. Note that $\mathbf{H}_t(\omega_t)$ is an $m \times m$ matrix with parameter ω_t. The (i,j) entry in the matrix $\mathbf{H}_t(\omega_t)$ is denoted as $\mathbf{H}_{t,\omega_t}(i,j)$.

Analogous to Theorem 1, the Nash stability of DM t is determined by Definition 3.

Definition 3. For an MOGM, state $s_i \in S$ is Nash stable for DM t iff ω_t satisfies $\mathbf{H}_{t,\omega_t}(i,j) \leq 0$ for all $j \neq i, j = 1, 2, \cdots, m$.

Definition 4. Fix $s_i \in S$. If s_i is Nash stable for DM t, then ω_t satisfies $\mathbf{H}_{t,\omega_t}(i,j) \leq 0$ for all $j \neq i, j = 1, 2, \cdots, m$. Moreover, for all $\omega_t, t \in N$, the area defined by the intersection $\Delta = \bigcap\limits_{\substack{t \in N \\ j \neq i}} \{\omega_t : \mathbf{H}_{t,\omega_t}(i,j) \leq 0\}$ is the location of the Nash equilibrium.

Note that for a MOGM, a state is Nash stable for a DM if that DM would not choose to move away from it. A Nash equilibrium of the MOGM is a state that is Nash stable for all DMs.

3 Application: US-China Trade Dispute

3.1 Background of US-China Trade Dispute

The trade dispute between the US and China, also known as the US-China trade dispute, is an ongoing economic conflict between the world's two largest national economies. It began on March 23, 2018, when the US imposed a tax on \$60 billion of Chinese imports.

China's large trade surplus with the US, China's non-compliance with WTO commitments, and China's tendency to disputed use of US technology have been suggested as reasons for the dispute. The underlying cause of the dispute is undoubtedly related to the intensification of domestic conflicts over the distribution of wealth in the US, the gradual decline of American hegemony, and China's rapid rise that seems to seriously threaten US interests. In addition, changes in the international situation will also have a huge impact on the trend of the trade dispute. Therefore, it would be best if the dispute could be resolved peacefully, both for the benefit of both sides and for global economic stability.

Recently, a high-level US and China government trade delegation reached a framework agreement to resolve the dispute. However, the implementation of the framework agreement requires structural adjustments to the bilateral economy, especially the Chinese economy, and this is difficult to achieve quickly. As a consequence, the US claimed that China did not fully implement the agreement, and the US began to impose more tariffs, causing China to retaliate. In this way, US and China cycled back and forth between imposing tariffs, reducing them, and then imposing them again. This process is called a "Thucydides trap" [24].

In view of the reasons for the trade dispute, and consistent with the preferences of the US and China, three scenarios are plausible for the dispute. They will be analyzed below, considering both short-term and long-term objectives. Based on the MOGM of Sect. 2, a scenario demonstration of the development path of the dispute can be conducted.

3.2   Multi-objective Graph Model of US-China Trade Dispute

The DMs in the US-China trade dispute are US and China. The US's goals are to obtain more favorable conditions for the US, and to reach a new agreement that the US President Trump called "don't lose money", rather than to fully implement trade controls and raise trade barriers. China's main goals are that the US recognize China's market economy status, establish an equal basis for negotiation, and gain a greater voice in the global economic and trade system. To secure their goals, each DM has two strategies - to impose tariffs or not. All possible combinations of DM's strategies are then examined to identify the states or situations in the dispute. Two DMs, their strategies, and states of the conflict are shown in Table 1.

Table 1. DMs, strategies and states for the US-China trade dispute

		China	
		Impose	Don't impose
US	Impose	s_1	s_2
	Don't impose	s_3	s_4

As in Table 1, the US is the row DM which controls the two row strategies of "Impose" by continually imposing tariffs on imports from China, or "Don't impose" by making concessions and stopping imposing tariffs. China is the column DM which also has two strategies "Impose" and "Don't impose". When each DM selects a strategy, a state is represented by a cell in the matrix. Each cell is assigned a state number as shown in Table 1. For instance, when the US chooses "Impose" and China selects "Don't impose", then state s_2 is formed as shown in the upper right-hand cell.

The short term o_1 and long term o_2 impact of the dispute on each country are considered as two objectives or dimensions. Let the weights of US and China on the two objectives be $\omega_{US} = (1 - \omega, \omega)$ and $\omega_C = (1 - \theta, \theta)$, respectively. The MOGM is used to analyze the bilateral trade dispute. Three scenarios for the dispute, depending on the preferences of two sides, are assumed as follows.

(1) In Scenario 1, either in the short or long term, both DMs prefer a state in which they impose tariffs and their opponent does not. The least preferred state for a DM is not to impose tariffs while the other DM does. In particular, the US prefers state s_2 to

state s_3 ($s_2 \succ s_3$) and China prefers state s_3 to state s_2 ($s_3 \succ s_2$) both in the short and long term. In the short term, no matter whether it is the US or China, the DM will prefers to impose tariffs rather than end the dispute ($s_1 \succ s_4$). However, in the long term, both DM realize that the dispute will have a negative impact on future economic development. Therefore, in this dimension, the two DMs prefer to not impose tariffs on each other rather than not ($s_4 \succ s_1$). In this scenario, the orders of preference of the two DMs according to the two objectives are shown in Table 2.

Table 2. The orders of preference of US and China in short and long term in Scenario 1

Scenario 1	US
Short term $(1 - \omega)$	$s_2 \succ s_1 \succ s_4 \succ s_3$
Long term (ω)	$s_2 \succ s_4 \succ s_1 \succ s_3$
China	
Short term $(1 - \theta)$	$s_3 \succ s_1 \succ s_4 \succ s_2$
Long term (θ)	$s_3 \succ s_4 \succ s_1 \succ s_2$

The UM matrices for US and China are as follows

$$
J_{US} = \begin{pmatrix} 0 & 0 & 1 & 0 \\ 0 & 0 & 0 & 1 \\ 1 & 0 & 0 & 0 \\ 0 & 1 & 0 & 0 \end{pmatrix}, \quad
J_C = \begin{pmatrix} 0 & 1 & 0 & 0 \\ 1 & 0 & 0 & 0 \\ 0 & 0 & 0 & 1 \\ 0 & 0 & 1 & 0 \end{pmatrix}.
$$

According to Definition 1, the preference matrices of US and China in the short and long term are

$$
P_{US}^S = \begin{pmatrix} 0 & 1 & -1 & -1 \\ -1 & 0 & -1 & -1 \\ 1 & 1 & 0 & 1 \\ 1 & 1 & -1 & 0 \end{pmatrix}, \quad
P_{US}^L = \begin{pmatrix} 0 & 1 & -1 & 1 \\ -1 & 0 & -1 & -1 \\ 1 & 1 & 0 & 1 \\ 1 & 1 & -1 & 0 \end{pmatrix},
$$

$$
P_C^S = \begin{pmatrix} 0 & -1 & 1 & -1 \\ 1 & 0 & 1 & 1 \\ -1 & -1 & 0 & -1 \\ 1 & -1 & 1 & 0 \end{pmatrix}, \quad
P_C^L = \begin{pmatrix} 0 & -1 & 1 & 1 \\ 1 & 0 & 1 & 1 \\ -1 & -1 & 0 & -1 \\ -1 & -1 & 1 & 0 \end{pmatrix}.
$$

Based on Eq. (1), UMP matrices for US and China can be calculated as follows

$$
H_{US}(\omega) = \begin{pmatrix} 0 & 0 & -1 & 0 \\ 0 & 0 & 0 & -1 \\ 1 & 0 & 0 & 0 \\ 0 & 1 & 0 & 0 \end{pmatrix}, \quad
H_C(\theta) = \begin{pmatrix} 0 & -1 & 0 & 0 \\ 1 & 0 & 0 & 0 \\ 0 & 0 & 0 & -1 \\ 0 & 0 & 1 & 0 \end{pmatrix}.
$$

According to Definition 4, we get one Nash equilibrium of the dispute in Scenario 1, which is shown in Fig. 1.

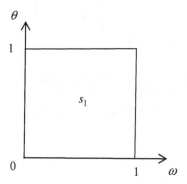

Fig. 1. Nash equilibrium in Scenario 1

Figure 1 shows that, in Scenario 1, no matter in which objective the two DMs operate under, the final Nash equilibrium is state s_1. That is, the two DMs will always be caught in a fierce battle of imposing tariffs on each other. In this circumstance, neither of them has any incentive to move away from state s_1.

(2) In Scenario 2, in the short term, the orders of preference of US and China are the same as in Scenario 1. In the long term, the difference between Scenario 2 and Scenario 1 is that the orders of preference of two DMs changes. They both regard state s_4 as their most preferred state, they don't want to impose tariffs on each other. Both sides uphold the concept of harmony. It means that in the long term the US changes its preference order from $s_2 \succ s_4$ in Scenario 1 to $s_4 \succ s_2$ in Scenario 2 and China could change its preference order from $s_3 \succ s_4$ in Scenario 1 to $s_4 \succ s_3$ in Scenario 2. The other preferences remain the same as in Scenario 1. In this Scenario, the orders of preference of the two DMs in the short and long term are shown in Table 3.

Table 3. The orders of preference of US and China in short and long term in Scenario 2

Scenario 2	US	
Short term $(1 - \omega)$	$s_2 \succ s_1 \succ s_4 \succ s_3$
Long term (ω)	$s_4 \succ s_2 \succ s_1 \succ s_3$
China	
Short term $(1 - \theta)$	$s_3 \succ s_1 \succ s_4 \succ s_2$
Long term (θ)	$s_4 \succ s_3 \succ s_1 \succ s_2$

Based on the Eq. (1), UMP matrices of US and China in Scenario 2 can be calculated as follows

$$\mathbf{H}_{US}(\omega) = \begin{pmatrix} 0 & 0 & -1 & 0 \\ 0 & 0 & 0 & 2\omega - 1 \\ 1 & 0 & 0 & 0 \\ 0 & 1 - 2\omega & 0 & 0 \end{pmatrix}, \mathbf{H}_C(\theta) = \begin{pmatrix} 0 & -1 & 0 & 0 \\ 1 & 0 & 0 & 0 \\ 0 & 0 & 0 & 2\theta - 1 \\ 0 & 0 & 1 - 2\theta & 0 \end{pmatrix}.$$

According to Definition 4, s_1 is a Nash equilibrium when $0 \le \omega \le 1$ and $0 \le \theta \le 1$, and s_4 is a Nash equilibrium iff $0.5 \le \omega \le 1$ and $0.5 \le \theta \le 1$. Thus, the Nash equilibria of the dispute in Scenario 2 can be shown in Fig. 2.

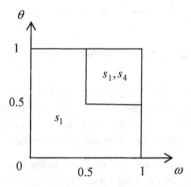

Fig. 2. Nash equilibrium in Scenario 2

Figure 2 indicates that, in Scenario 2, both DMs may change their orders of preference, that is, both DMs may make concessions in the negotiations, and state s_4 without tariffs is likely to be a Nash equilibrium. Why is it possible, not certain? Because, as shown in Fig. 2, state s_4 is an equilibrium solution only when both DMs focus more on the long term, and state s_1 is also an equilibrium solution at this time.

(3) In Scenario 3, in the short term, the orders of preference of the US and China are the same as in Scenario 1 and 2. In the long term, the two DMs change further. They both regard state s_1 as their least preferred state, they don't want to impose tariffs on each other. Why do DMs make such changes? The two DMs have repeatedly imposed tariffs, and the damage and losses to the domestic economy have exceeded their capacity. Then Scenario 3 will occur. In the long term, the US changes its preference from $s_1 \succ s_3$ in Scenario 2 to $s_3 \succ s_1$ in Scenario 3 and China changes its preference from $s_1 \succ s_2$ in Scenario 2 to $s_2 \succ s_1$ in Scenario 3. The other preferences remain the same as Scenario 2. In this Scenario, the orders of preference of the two DMs in the short and long term are shown in Table 4.

Table 4. The orders of preference of US and China in short and long term in Scenario 3

Scenario 3	US
Short term $(1 - \omega)$	$s_2 \succ s_1 \succ s_4 \succ s_3$
Long term (ω)	$s_4 \succ s_2 \succ s_3 \succ s_1$
China	
Short term $(1 - \theta)$	$s_3 \succ s_1 \succ s_4 \succ s_2$
Long term (θ)	$s_4 \succ s_3 \succ s_2 \succ s_1$

Based on the Eq. (1), UMP matrices of US and China in Scenario 3 can be cal-culated as follows

$$\mathbf{H}_{US}(\omega) = \begin{pmatrix} 0 & 0 & 2\omega - 1 & 0 \\ 0 & 0 & 0 & 2\omega - 1 \\ 1 - 2\omega & 0 & 0 & 0 \\ 0 & 1 - 2\omega & 0 & 0 \end{pmatrix},$$

$$\mathbf{H}_{C}(\theta) = \begin{pmatrix} 0 & 2\theta - 1 & 0 & 0 \\ 1 - 2\theta & 0 & 0 & 0 \\ 0 & 0 & 0 & 2\theta - 1 \\ 0 & 0 & 1 - 2\theta & 0 \end{pmatrix}.$$

According to Definition 4, s_1 is a Nash equilibrium when $0 \leq \omega \leq 0.5$ and $0 \leq \theta \leq 0.5$, s_2 is a Nash equilibrium when $0 \leq \omega \leq 0.5$ and $0.5 \leq \theta \leq 1$, s_3 is a Nash equilibrium when $0.5 \leq \omega \leq 1$ and $0 \leq \theta \leq 0.5$, and s_4 is a Nash equilibrium iff $0.5 \leq \omega \leq 1$ and $0.5 \leq \theta \leq 1$. Thus, Nash equilibria of the dispute in Scenario 3 are shown in Fig. 3.

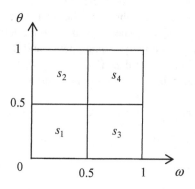

Fig. 3. Nash equilibrium in Scenario 3

The results of Scenario 3 have the following implications. First, the bilateral trade conflict can be stable in a peaceful state s_4 only if both DMs regard "Don't impose tariffs" as their preferred state and "Impose tariffs" on each other as the least desirable state. Second, if the weights ω and θ are large enough, that is, both sides pay enough attention to the long term interests, then peace state s_4 will be stable. Third, if one or both DMs are dissatisfied with the current peace state, s_4, concessions made by the opponent, or if the opponent does not implement the negotiation conditions in accordance with expectations, the current stable state may become unbalanced and the conflict enters into a state of struggle.

4 Multi-objective Graph Model with Interval Preference Weights

Crisp preference represents a certain (definite) preference between two states. For example, if the order of states for DM t is $s_1 \succ s_2 \succ s_3 \succ s_4$, then the preference matrix is

$$
\mathbf{P}_t = \begin{pmatrix} 0 & -1 & -1 & -1 \\ 1 & 0 & -1 & -1 \\ 1 & 1 & 0 & -1 \\ 1 & 1 & 1 & 0 \end{pmatrix}.
$$

Apparently, in this case, the degree or strength of preference of s_1 over s_3 is greater than the strength of preference of s_1 over s_2. However, in the structure of crisp preference, all degrees of preference are equal to one. In other words, crisp preference cannot express the degree or strength of preference between two states. Furthermore, in real management situations, there is commonly a great deal of fuzziness. DMs are often unclear or uncertain about their preference between two states for various reasons, such as cultural or educational factors, personal habits, lack of information, or the inherent vagueness of human judgment. For these two reasons, an interval number may be the most suitable data expression to describe preference.

An interval [25] is a special subset of the real number set $\Re$, denoted by $\hat{a} = [a, \bar{a}] = \{x \in \Re | a \leq x \leq \bar{a}\}$, where a and $\bar{a}$ are the left endpoint and the right endpoint of the interval $\hat{a}$, respectively. Sometimes a and $\bar{a}$ are called the lower and upper limits or bounds of the interval $\hat{a}$. If $\bar{a} \geq a \geq 0$, then $\hat{a}$ is called a positive interval number. In particular, if $a = \bar{a}$, then the interval number $\hat{a}$ reduces to a real number.

The basic arithmetic operations for intervals are defined as follows [25, 26].

Definition 5. Let $\hat{a} = [a, \bar{a}]$ and $\hat{b} = [b, \bar{b}]$ be two intervals, and let λ be a real number. Then

(1) $\hat{a} + \hat{b} = [a + b, \bar{a} + \bar{b}]$;
(2) $\lambda \hat{a} = \begin{cases} [\lambda a, \lambda \bar{a}], & if\ \lambda \geq 0 \\ [\lambda \bar{a}, \lambda a], & if\ \lambda < 0 \end{cases}$.

Definition 6. For a graph model G, the interval preference matrix for DM t is an $m \times m$ matrix, $\hat{\mathbf{P}}_t$ with (s_i, s_j) entries

$$\hat{\mathbf{P}}_t(s_i, s_j) = \begin{cases} \hat{a}, & \text{if } s_j \succ_t s_i, \\ 0, & \text{if } s_j \sim_t s_i, \\ -\hat{a}, & \text{if } s_j \prec_t s_i, \end{cases}$$

where $\hat{a}$ is a positive interval number. The preferences of DM t for s_i over s_j are represented by intervals. A value $\hat{\mathbf{P}}_t(s_i, s_j) = \hat{a} = [a, \bar{a}]$ in the interval preference matrix indicates the degree or strength of the preference for s_i over s_j for DM t. $\hat{\mathbf{P}}_t(s_i, s_j) = 0$ indicates that DM t is indifferent between s_i and s_j, while $\hat{\mathbf{P}}_t(s_i, s_j) = -\hat{a} = [-\bar{a}, -a]$ implies that DM t prefers state s_j to state s_i.

The interval preference matrix for DM t on an objective o_k, $k = 1, 2, \cdots, K$ is denoted by a matrix $\hat{\mathbf{P}}_{tk} = ([\underline{\mathbf{P}}_{tk}, \bar{\mathbf{P}}_{tk}])_{m \times m}$. In order to incorporate all objectives, the comprehensive preference matrix $\hat{\mathbf{P}}_t(\omega_t)$ with parameter ω_t for DM t can be calculated by

$$\hat{\mathbf{P}}_t(\omega_t) = \sum_{k=1}^{K} \omega_{tk} \hat{\mathbf{P}}_{tk}. \tag{4}$$

Then, for a MOGM, the UM matrix including interval preference weights (UMIP) for DM t can be calculated by

$$\hat{\mathbf{H}}_t(\omega_t) = \mathbf{J}_t \circ \hat{\mathbf{P}}_t(\omega_t), \tag{5}$$

where "$\circ$" denotes the Hadamard product. Note that $\hat{\mathbf{H}}_t(\omega_t)$ is an $m \times m$ matrix with parameter ω_t. The (i, j) entry in the matrix $\hat{\mathbf{H}}_t(\omega_t)$ is $\hat{\mathbf{H}}_{t,\omega_t}(i,j) = [\underline{\mathbf{H}}_{t,\omega_t}(i,j), \bar{\mathbf{H}}_{t,\omega_t}(i,j)]$.

The logical definition of Nash stability of the MOGM with interval preference weights is given as follows.

Definition 7. For MOGM with interval preference weights, state $s_i \in S$ is interval Nash stable for DM t iff ω_t satisfies $\hat{\mathbf{H}}_{t,\omega_t}(i, j) \leq 0$ for all $j \neq i, j = 1, 2, \cdots, m$.

Since $\hat{\mathbf{H}}_{t,\omega_t}(i, j)$ is an interval number, one can obtain different results on $\hat{\mathbf{H}}_{t,\omega_t}(i, j) = [\underline{\mathbf{H}}_{t,\omega_t}(i, j), \bar{\mathbf{H}}_{t,\omega_t}(i, j)] \leq [0, 0]$ by using different ranking methods on interval numbers. In this paper, assume that $\hat{\mathbf{H}}_{t,\omega_t}(i,j) \leq 0$ iff $\underline{\mathbf{H}}_{t,\omega_t}(i, j) \leq \bar{\mathbf{H}}_{t,\omega_t}(i, j) \leq 0$. Thus, the following Definition 8 on interval Nash equilibrium is proposed for MOGM with interval preference weights.

Definition 8. Fix $s_i \in S$. If s_i is interval Nash stable for DM t, then interval preference weight ω_t satisfies $\bar{\mathbf{H}}_{t,\omega_t}(i,j) \leq 0$ for all $j \neq i, j = 1, 2, \cdots, m$. Moreover, for all $\omega_t, t \in N$, the area defined by the intersection $\Delta = \bigcap_{\substack{t \in N \\ j \neq i}} \{\omega_t : \bar{\mathbf{H}}_{t,\omega_t}(i,j) \leq 0\}$ is the location of the Nash equilibrium.

5 US-China Trade Dispute with Interval Preference

We analyze Scenario 3 as an example. For this case, using Definition 3 and Definition 4, each preference relation can be characterized by a degree or strength of preference and a relative degree or intensity of preference as shown in Table 5.

Table 5. Interval preference of four states relation

Preference structure	Interval preference degree
≻≻≻	$[0.8, 1]$
≻≻	$[0.6, 0.8]$
≻	$[0.2, 0.6]$
∼	$[0, 0]$
≺	$[-0.6, -0.2]$
≺≺	$[-0.8, -0.6]$
≺≺≺	$[-1, -0.8]$

Then, the interval preference matrices for the US and China in the short and long term are

$$\hat{\mathbf{P}}_{US}^{S} = \begin{pmatrix} [0,0] & [0.2,0.6] & [-0.8,-0.6] & [-0.6,-0.2] \\ [-0.6,-0.2] & [0,0] & [-1,-0.8] & [-0.8,-0.6] \\ [0.6,0.8] & [0.8,1] & [0,0] & [0.2,0.6] \\ [0.2,0.6] & [0.6,0.8] & [-0.6,-0.2] & [0,0] \end{pmatrix},$$

$$\hat{\mathbf{P}}_{US}^{L} = \begin{pmatrix} [0,0] & [0.6,0.8] & [0.2,0.6] & [0.8,1] \\ [-0.8,-0.6] & [0,0] & [-0.6,-0.2] & [0.2,0.6] \\ [-0.6,-0.2] & [0.2,0.6] & [0,0] & [0.6,0.8] \\ [-1,-0.8] & [-0.6,-0.2] & [-0.8,-0.6] & [0,0] \end{pmatrix},$$

$$\hat{\mathbf{P}}_{C}^{S} = \begin{pmatrix} [0,0] & [-0.8,-0.6] & [0.2,0.6] & [-0.6,-0.2] \\ [0.6,0.8] & [0,0] & [0.8,1] & [0.2,0.6] \\ [-0.6,-0.2] & [-1,-0.8] & [0,0] & [-0.8,-0.6] \\ [0.2,0.6] & [-0.6,-0.2] & [0.6,0.8] & [0,0] \end{pmatrix},$$

$$\hat{\mathbf{P}}_{C}^{L} = \begin{pmatrix} [0,0] & [0.2,0.6] & [0.6,0.8] & [0.8,1] \\ [-0.6,-0.2] & [0,0] & [0.2,0.6] & [0.6,0.8] \\ [-0.8,-0.6] & [-0.6,-0.2] & [0,0] & [0.2,0.6] \\ [-1,-0.8] & [-0.8,-0.6] & [-0.6,-0.2] & [0,0] \end{pmatrix}.$$

Based on Eq. (5), UMIP matrices for the US and China in Scenario 3 can be calculated as follows

$$\hat{\mathbf{H}}_{US}(\omega) = \begin{pmatrix} 0 & 0 & [\omega - 0.8, 1.2\omega - 0.6] & 0 \\ 0 & 0 & 0 & [\omega - 0.8, 1.2\omega - 0.6] \\ [-1.2\omega + 0.6, -\omega + 0.8] & 0 & 0 & 0 \\ 0 & [-1.2\omega + 0.6, -\omega + 0.8] & 0 & 0 \end{pmatrix},$$

$$\hat{\mathbf{H}}_{C}(\theta) = \begin{pmatrix} 0 & [\theta - 0.8, 1.2\theta - 0.6] & 0 & 0 \\ [-1.2\theta + 0.6, -\theta + 0.8] & 0 & 0 & 0 \\ 0 & 0 & 0 & [\theta - 0.8, 1.2\theta - 0.6] \\ 0 & 0 & [-1.2\theta + 0.6, -\theta + 0.8] & 0 \end{pmatrix}.$$

According to Definition 8, s_1 is an interval Nash equilibrium when $0 \leq \omega \leq 0.5$ and $0 \leq \theta \leq 0.5$, s_2 is an interval Nash equilibrium when $0 \leq \omega \leq 0.5$ and $0.8 \leq \theta \leq 1$, s_3 is an interval Nash equilibrium when $0.8 \leq \omega \leq 1$ and $0 \leq \theta \leq 0.5$, and s_4 is an interval Nash equilibrium iff $0.8 \leq \omega \leq 1$ and $0.8 \leq \theta \leq 1$. Thus, the interval Nash equilibria of the dispute in Scenario 3 can be shown in Fig. 4.

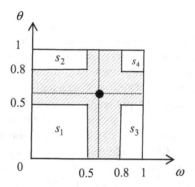

Fig. 4. Interval Nash equilibrium in Scenario 3

There is a shaded hole in Fig. 4. The hole means that it is uncertain if the Nash stability s_1 is going to become s_2 with a change from short term to long term for China. For example, at the point $(0.6, 0.6)$, $\hat{\mathbf{H}}_{US}(1, 3) = [-0.2, 0.12]$ indicates the degree or strength of the US is likely to leave state s_1 to state s_3. If the US uses the average of interval number, then US won't stay in state s_1, US would change its strategy to state s_3 since $\mathbf{H}_{US}(1, 3) = -0.04$. This result is consistent with our idea of introducing an uncertain preference. This kind of uncertainty is more suitable for real world disputes. It is indeed a challenging problem in this uncertain environment.

From the above, we get the following management enlightenment for this kind of bilateral trade disputes.

(1) This shows that whether US and China can stabilize in the peace state s_4 for a long time mainly depends on whether the two sides can reasonably take a long term view, not be tempted by the immediate short term, and no longer adopt

sophisticated sanctions and counter sanctions such as deterrence and temptation. It is obvious that achieving win-win cooperation between the US and China in both economy and trade is not only the rational choice for China but also for the US.

(2) Any unilateral concession won't reverse the stable state of the dispute. The most ideal state is that the two DMs, through peaceful consultations, focus on the long term, make concessions, and maintain stability.

(3) No matter how sophisticated sanctions and anti-sanction strategies are, if either of the DMs in the dispute lacks the goodwill to cooperate and promote mutual well-being and prosperity, then both DMs of the dispute are likely to fall into a swirl of fierce fighting, which will eventually lead to the undesirable result of both losing.

(4) Even if the "Thucydides trap" is unavoidable, we still need to analysis and comprehend the situation and understand that mutual concessions and cooperation are the paths to prevent economic and social decline.

6 Conclusion

A MOGM is defined in this paper to incorporate each DM's objectives. Mathematical matrix representations of preference are introduced by using values $1, 0$, and -1. It can express three concepts of preference by DM: positive preference, indifference, and negative preference in one matrix. The comprehensive preference matrix with weight parameters on objectives and the UMP preference for MOGM is developed. Furthermore, the subjective uncertainty of DMs is considered. Interval numbers are used to express the degree of uncertainty of preference. Subsequently, Nash equilibrium and interval Nash equilibrium solution methods are developed for MOGMs and the dependence of equilibria on weights of objectives is also shown. A detailed modeling and calculation process in US-China trade dispute is also explained and demonstrated. The implications and in-depth result analysis for DMs have been given. The MOGM provides DMs with guidance for how to act strategically in bilateral trade disputes that occur in the real world.

Further studies can be carried out within MOGM. A preference structure to incorporate random uncertainty and probability deserves further research. Nash stability for MOGM is developed in this paper. For other classical stabilities for MOGM, such as GMR, SMR and SEQ, the definitions and the solution methods still need to be studied. Furthermore, MOGM could be expanded by taking into account coalitions among DMs.

References

1. Blackwell, O.: An analog of the minimax theorem for vector payoffs. Pac. J. Math. **6**(1), 1–8 (1956)
2. Shapley, L.S.: Equilibrium points in games with vector payoff. Naval Res. Logist. Q. **6**(1), 57–61 (1959)
3. Zeleny, M.: Games with multiple payoffs. Int. J. Game Theory **5**(4), 252 (1976)

4. Borm, P.E.M., Tijs, S.H., Aarssen, J.C.M.: Pareto equilibria in multiobjective games. Methods Oper. Res. **60**, 303–312 (1988)
5. Charnes, A., Huang, Z.M., Rousseau, J.J., Wei, Q.L.: Corn extremal solutions of multi-payoff games with cross-constrained strategy set. Optimization **21**(1), 51–69 (1990)
6. Zhao, J.: The equilibria of a multiple objective game. Int. J. Game Theory **20**(2), 171–182 (1999). https://doi.org/10.1007/BF01240277
7. Yu, J., Yuan, X.Z.: The study of Pareto equilibria for multiobjective games by fixed point and Ky Fan minimax inequality methods. Comput. Math. Appl. **35**(9), 17–24 (1998)
8. Nishizaki, I., Sakawa, M.: Fuzzy and Multiobjective Games for Conflict Resolution. Physica, Heidelberg (2001). https://doi.org/10.1007/978-3-7908-1830-7
9. Fang, L., Hipel, K.W., Kilgour, D.M.: Interactive Decision Making: The Graph Model for Conflict Resolution. Wiley, New York (1993)
10. Kilgour, D.M., Hipel, K.W.: Conflict analysis methods: the graph model for conflict resolution. In: Kilgour, D., Eden, C. (eds.) Handbook of Group Decision and Negotiation. AGDN, vol. 4, pp. 203–222. Springer, Dordrecht (2010). https://doi.org/10.1007/978-90-481-9097-3_13
11. Xu, H., Hipel, K.W., Kilgour, D.M., Fang, L.: Conflict Resolution Using the Graph Model: Strategic Interactions in Competition and Cooperation. Springer, Cham (2018). https://doi.org/10.1007/978-3-319-77670-5
12. Xu, H., Kilgour, D.M., Hipel, K.W.: Matrix representation of solution concepts in graph models for two decision-makers with preference uncertainty. Dyn. Contin. Discrete Impuls. Syst. **14**, 703–707 (2007)
13. Xu, H., Kilgour, D.M., Hipel, K.W.: Matrix representation of conflict resolution in multiple-decision-maker graph models with preference uncertainty. Group Decis. Negot. **20**(6), 755–779 (2011). https://doi.org/10.1007/s10726-010-9188-4
14. Fang, L., Hipel, K.W., Kilgour, D.M., Peng, X.: A decision support system for interactive decision making, part 1: model formulation. IEEE Trans. Syst. Man Cybern. Part C **SMC-33**(1), 42–55 (2003)
15. Fang, L., Hipel, K.W., Kilgour, D.M., Peng, X.: A decision support system for interactive decision making, part 2: analysis and output interpretation. IEEE Trans. Syst. Man Cybern. Part C **SMC-33**(1), 56–66 (2003)
16. Li, K.W., Hipel, K.W., Kilgour, D.M., et al.: Preference uncertainty in the graph model for conflict resolution. IEEE Trans. Syst. Man Cybern. - Part A: Syst. Hum. **34**(4), 507–520 (2004)
17. Bashar, M.A., Kilgour, D.M., Hipel, K.W.: Fuzzy preferences in the graph model for conflict resolution. IEEE Trans. Fuzzy Syst. **20**(4), 760–770 (2012)
18. Hipel, K.W., Kilgour, D.M., Bashar, M.A.: Fuzzy preferences in multiple participant decision making. Sci. Iran. **18**(3), 627–638 (2011)
19. Ke, Y., Fu, B., De, M., Hipel, K.W.: A hierarchical multiple criteria model for eliciting relative preferences in conflict situations. J. Syst. Sci. Syst. Eng. **21**(1), 56–76 (2012)
20. Ke, Y., Li, K.W., Hipel, K.W.: An integrated multiple criteria preference ranking approach to the Canadian west coast port congestion problem. Expert Syst. Appl. **39**(10), 9181–9190 (2012)
21. Kuang, H., Bashar, M.A., Hipel, K.W., et al.: Grey-based preference in a graph model for conflict resolution with multiple decision makers. IEEE Trans. Syst. Man Cybern.: Syst. **45**(9), 1254–1267 (2015)
22. Rego, L.C., Santos, A.M.: Probabilistic preferences in the graph model for conflict resolution. IEEE Trans. Syst. Man Cybern.: Syst. **45**(4), 595–608 (2015)
23. Nash, J.: Non-cooperative games. Ann. Math. **54**(2), 286–295 (1951)

24. Moosa, I.A.: The Thucydides trap as an alternative explanation for the US-China trade war. Global J. Emerg. Market Econ. **12**, 42–55 (2020)
25. Moore, R.E.: Method and Application of Interval Analysis. SIAM, Philadelphia (1979)
26. Li, D., Nan, J., Zhang, M.: Interval programming models for matrix games with interval payoffs. Optimiz. Methods Softw. **27**(1), 1–16 (2012)

A Novel Conflict Resolution Model Based on the Composition of Probabilistic Preferences

Annibal P. Sant'anna[1], Ana Paula C. S. Costa[2],
and Maisa M. Silva[2(✉)]

[1] Department of Management Engineering, Escola de Engenharia,
Universidade Federal Fluminense – UFF, Rua Passos da Pátria,
156, São Domingos, Niterói, RJ 24210-240, Brazil
annibal.parracho@gmail.com
[2] Department of Management Engineering, Universidade Federal
de Pernambuco – UFPE, Caixa Postal 5125, Recife, PE 52070-970, Brazil
apcabral@hotmail.com, maisa@cdsid.org.br

Abstract. The purpose of this paper is to develop a four-stage conflict resolution model. In the first stage, a multicriteria model is developed for each of the conflicting parties, taken as decision makers (DMs) facing evaluations of a set of alternatives according to proper criteria. In the second stage, the composition of probabilistic preferences (CPP) methodology is applied to identify the best alternative for each of the conflicting parties. In the third stage, negotiation is carried out to remove alternatives and to focus on the subset of best alternatives for the group of DMs. The fourth stage consists of applying CPP again to choose one among the remaining alternatives. The model is illustrated by means of applying it to a real-world conflict in Brazil, related to implementation of the New Recife Project. The main features of the model are that it allows the DMs (i) to understand differences and proximities among the positions of each of them, (ii) to strategically reduce the initial set of alternatives, (iii) to advance in their positions towards a common goal, and (iv) to construct a unique final solution quickly.

Keywords: Composition of Probabilistic Preferences (CPP) · Conflict Resolution Model · Negotiation · Multicriteria

1 Introduction

The occurrence of conflicts, which are situations in which the actors involved present different perceptions and/or preferences, is very common within group decision and negotiation processes. The literature presents different methodologies to analyze this type of situation, among which are: Game Theory [1], Metagame Analysis [2], Conflict Analysis [3], and the Graph Model for Conflict Resolution (GMCR) [4, 5].

In this sense, the present paper advances in this area by proposing a methodology of analysis and resolution of conflicts, named Conflict Resolution Model with Composition of Probabilistic Preferences (CRMCPP), in which the decision process is divided

© Springer Nature Switzerland AG 2020
D. C. Morais et al. (Eds.): GDN 2020, LNBIP 388, pp. 21–31, 2020.
https://doi.org/10.1007/978-3-030-48641-9_2

into successive stages. Throughout this process, the Composition of Probabilistic Preferences (CPP) methodology [6] is applied to guide the evolution of the negotiators' positions.

The authors of [7] used CPP to deal with the presence of uncertainty in the assessment of preferences. Here, differently, CPP composition rules, based on the weighting of the criteria by the different negotiators, or other forms of joint evaluation, are used in the group decision context [8, 9]. The solution can be defined by mutual agreement to determine an alternative that maximizes the joint preference. This may become difficult if there is disagreement among negotiators as to the positive or negative sense to be assigned to evaluations by one or more criteria. If this occurs, initially, guidelines to simplify positions are generated by applying CPP. Later, if negotiators do not reach full agreement, an automatic final solution is offered by re-applying CPP.

The main feature of CRMCPP is its ability to aid negotiators to understand differences and proximities between each other's positions, thereby allowing them to advance in their positions and to construct a final solution quickly. As a practical implication, the model proposed can be used in different real situations where a group of negotiators face difficulties in finding a consensual solution. The originality of this proposal is to simplify and accelerate the path of consensus by applying CPP.

This paper is structured as follows. Section 2 presents the CPP methodology and Sect. 3 the model of conflict resolution proposed using CPP. In Sect. 4, one real-world application of CRMCPP is presented using the data presented in [7]. Section 5 presents final considerations on the proposal, draws some conclusions and makes suggestions for future lines of research.

2 Composition of Probabilistic Preferences (CPP)

CPP is a methodology designed to take into account, in the composition of decision criteria, the presence of imprecision in the assessments of preference. By treating the evaluations as observations of random variables, it generates rules based on probabilities of choice to rank alternatives which are evaluated by different criteria or different experts. The first stage of applying CPP consists of associating a probability distribution to each evaluation of each alternative according to each criterion. The key idea is that each evaluation assessment determines a value around which it may vary. Instead of exact values, the evaluations are treated as location parameters of the probability distributions of possible values that would be assigned to the same alternative in other preference assessments under similar circumstances. Thus, the exact values of the decision matrix are turned into parameters of random variables.

On comparing these distributions, the probability that each alternative is the most preferred according to each criterion is accessed. This probability of an alternative being the most preferred is calculated as the joint probability of the set of multivariate evaluations for which such an alternative presents an evaluation higher than any other. To present this concept formally, let $(a_{ij}, \ldots, a_{nj})$ denote a vector of evaluations of n alternatives by a criterion j and let X_{kj} denote a random variable with the distribution of preference for the alternative k according to criterion j. For any k, a_{kj} will be used as a location parameter for the distribution of X_{kj}. The probability of alternative i being the best according to criterion j is given by the integral for x varying along the domain of X_{ij} of P [$X_{kj} < $ x, for all $k \neq i$] with respect to the density of X_{ij}. Denoting by f_i this density of the distribution of the evaluation of alternative i by criterion j and denoting by F_{-i} the cumulative distribution function of the joint evaluations by criterion j of the $n - 1$ alternatives different from i, the probability of alternative i being the most preferred by criterion j is given by

$$M_{ij} = \int F_{-i}(x) f_i(x) dx \qquad (1)$$

On the other hand, knowing the probabilities of being the least preferred is also useful. They are given by

$$m_{ij} = \int (1 - F_{-i}(x)) f_i(x) dx \qquad (2)$$

Applying proper composition rules from the probabilities of preference relative to each criterion may derive global probabilities of preference. The most employed composition rule explores the basic concepts of conditional probabilities. The probabilities of being the best alternative according to each criterion are then considered as conditional on the preference for such a criterion. Thus, an unconditional global preference is determined as a linear combination of these conditional probabilities. The preferences for the criteria enter as the marginal probabilities of the conditioning events that constitute the weights of the linear combination. The determination of these weights is difficult as interaction among criteria may have to be taken into account.

Simpler forms of composition consist of using probabilities of unions and intersections which are also permitted because of the probabilistic transformation. To use these forms of composition, instead of assigning weights to the criteria, the DM is asked only to choose between a conservative and a progressive point of view and between an optimistic and a pessimistic point of view. In the first choice, the progressive DM wants to decide on the basis of probabilities of maximizing the preference according to the criteria, while the conservative DM prefers to consider the probabilities of not minimizing it. The progressive DM pays attention to distances to the extremes of excellence, while the conservative one pays attention to distances to the extremes of worst performance. The term conservative is associated with risk aversion, while the term progressive refers to a DM who is willing to take risks in order to achieve a higher standard of excellence.

On the other hand, in the optimism versus pessimism choice, the optimistic point of view considers satisfactory being the best in at least one criterion or not being the worst in any of them. The global score is then determined by the probability of maximizing preference according to at least one among the multiple criteria or not minimizing preference according to all of them. Alternatively, from the pessimistic point of view, the global preference is measured by the probability of maximizing the preference according to all the criteria or not minimizing it with respect to any of them. The expressions optimistic and pessimistic are related to the idea of believing that the most favorable, or least favorable, criterion will prevail, respectively. Consequently, by combining these positions, four measures are generated:

$$\text{Optimistic and conservative}: OC_i = 1 - \prod_j m_{ij} \tag{3}$$

$$\text{Optimistic and progressive}: OP_i = 1 - \prod_j (1 - M_{ij}) \tag{4}$$

$$\text{Pessimistic and conservative}: PC_i = \prod_j (1 - m_{ij}) \tag{5}$$

$$\text{Pessimistic and progressive}: PP_i = \prod_j M_{ij} \tag{6}$$

3 Conflict Resolution Model Based on Composition of Probabilistic Preferences (CRMCPP)

Two characteristics of CPP are explored in this paper with a view to resolving conflicts. First, there is the characteristic of implicitly assigning greater importance to the criteria with the greatest power to discriminate an option as the most preferred. Whatever the form of the multiple criteria used, the transformation of the initial evaluations into probabilities of being the most preferred increases the importance of those criteria that are most able to highlight a preferred option.

The second characteristic stems from the attention that CPP gives to the set of alternatives that are being compared in order to determine the context of the decision problem to be solved. As DMs who are also the negotiators reach agreement on which criteria adequately represent the opposing points of view, CPP can be used to help each negotiator eliminate the least preferred alternatives. By limiting the set of alternatives to those most preferred by at least one of the negotiators, the positions in conflict can be more clearly shown. Thus, it becomes easier for each negotiator to associate himself/herself with one or more of the others negotiators to reach a final decision in a space of simpler alternatives.

The decision problem is placed in terms of the search for an alternative that maximizes the joint preference with respect to the criteria maintained until the end. The alternative finally chosen is that one seen as providing the highest satisfaction not for a particular participant in the conflict but for the whole group. The group interest is represented by the set of criteria considered relevant by the end of the negotiation.

CPP is initially used to indicate to each negotiator a single alternative that best represents his/her own preferences. By examining these representative alternatives of the various positions, negotiators can identify close alternatives that, if they are kept competing with each other, will lessen the likelihood that each one of them will eventually be chosen. If, on the other hand, negotiators give up their own representatives and adhere to only one alternative to represent multiple negotiators, this will increase the probability that this will be the most preferred alternative in the last stage when CPP is applied for the final time. This encourages everybody to contribute towards simplifying the problem.

Any approach to the acceptance of the alternative that maximizes composite preference requires a preliminary stage of clear preference elicitation and negotiation that allows a set of preferred alternatives and preference criteria to be identified that are representative of the different interests. Therefore, a model for each negotiator is developed in the first stage. In the second stage, CPP is applied to each negotiator's criteria so as to identify an alternative that best represents him/her. In the third stage, the negotiators have the opportunity to unite, choosing only one among alternatives close to each other. In the last stage, CPP is applied to the set of remaining alternatives. Thus, CRMCPP is developed in four stages, as shown in Fig. 1.

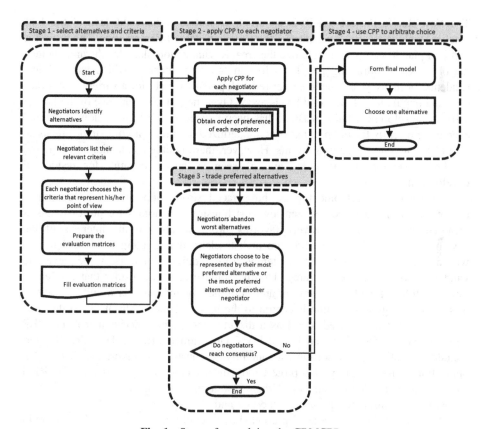

Fig. 1. Stages for applying the CRMCPP

In the second stage, the conservative pessimistic composition for each negotiator is applied, leading to the global evaluation by the joint probability of not minimizing the preference according to all the criteria in the model. This approach follows from the purpose of this stage is to identify an alternative that will represent the negotiator in the next stages. This alternative should have the lowest probability of misrepresentation in any aspect considered relevant by the negotiator.

In the fourth stage, considering that, if divergences remained until this point, it is impossible to fully satisfy all the negotiators, the form of combination used applies the optimistic and progressive point of view. The difficulty of reaching simultaneous satisfaction of conflicting criteria leads to the optimistic approach, in which one seeks to maximize the preference for at least one criterion, without establishing which one. And, assuming satisfaction maximization as the final goal, the conservative approach is replaced by the progressive one.

4 Numerical Example

The New Recife Project (NRP) is an urban project to reshape the area of the Jose Estelita Dock, in Recife, the capital of the Brazilian state of Pernambuco. The project is part of a plan to renew and revitalize the area. Revitalization plans for the area include the objective of attracting people to live in the area, most of which has had few residents for many years due to earlier redevelopments which, for a variety of reasons, drew the population of all social levels away. Over the last 20 years, one of the largest software development hubs in Brazil has been created in the historic downtown and even more recently the old port has been modernized both for sea-going trade and as an area of leisure and for cultural events. However, there is little residential housing, and so the conversion of the Dock area is seen as an opportunity for residential development.

Nevertheless, development plans have faced strong opposition for three main reasons. First, the balance between low-cost housing and homes for professionals is regarded by some as excessively favoring professionals and the interests of the developers. Secondly, some environmentalists claim that erecting residential towers will form a curtain that will cause the temperature of downtown to rise as the prevailing wind will no longer cool the area. Thirdly, some urbanists consider that residential towers will block landscape views of and from the city and that this would be a severe loss of an intangible asset that is central to the image of the sea-front of Recife.

The conflict is modeled in [7] as a dispute among three decision makers: NRP Support, NRP Opposition, and Recife Local Government (RLG). The NRP Support includes essentially a consortium of four private construction enterprises. The NRP opposition consists mainly of a protest movement called Occupy Estelita. The RLG includes the City Council and the Urban Development Council, a group consisting primarily of counselors representing the city and civil society.

The states of this conflict are represented by a set of four criteria, which measure in economic, environmental, social and political terms the impact expected. These four criteria have been determined by a group of six experts hired by RLG to evaluate the project. Two criteria (economic and political impacts) are measured in positive terms. In other words, they measure the benefits assuming that there is a possibility of the initiative being carried forward. The other two criteria measure the negative impacts of the initiative, should it be carried forward. The economic impact criterion is given by the approximate number of jobs created. The idea is that instead of making decisions unaided, the RLG decides to apply the conflict resolution technique presented in this paper. The four measures of impact are taken as decision criteria by all three negotiators. Initially, each negotiator compares the alternatives. In the final stage, the choice that is made is among the best alternatives for each negotiator.

It is assumed that the NRP consortium will choose one from three lines of action: proceed to complete the full NRP, proceed to finish the NRP with minor modifications, or proceed to finish the NRP with major modifications. The NRP Opposition may choose between not offering legal resistance or proceeding with a legal action against the NPR; and the RLG has four options, which are: support completion of full NRP, support completion of NRP with minor modifications, support completion of NRP with major modifications, and support suspension of the NRP. As the DMs cannot take their different actions simultaneously, this leads to a total of 24 combinations. A 25th alternative is given by the *status quo*, with the project being fully developed, RLG selecting no option and no legal action being presented by the NRP opposition.

In Table 1, the three first rows correspond to the three options available to the consortium: full project, project with minor changes or project with major changes. The fourth row presents the only available option for the NRP Opposition which is to propose a legal action to prevent the construction of the project. In the four last rows, the options of the RLG are presented (full project, project with minor or major changes or the suspension of the project. In Table 1, a letter Y in the rows is the situation in which the option is chosen by the DM and the letter N represents the situation in which the option is not chosen by the DM.

Table 1. Options and feasible states in the NRP Conflict

	1	2	3	4	5	6	7	8	9	10	11	12	13
1.1 Full	Y	Y	N	N	Y	N	N	Y	N	N	Y	N	N
1.2 Minor change	N	N	Y	N	N	Y	N	N	Y	N	N	Y	N
1.3 Major change	N	N	N	Y	N	N	Y	N	N	Y	N	N	Y
2. Legal action	N	N	N	N	Y	Y	Y	N	N	N	Y	Y	Y
3.1 Full support	N	Y	Y	Y	Y	Y	Y	N	N	N	N	N	N
3.2 Minor support	N	N	N	N	N	N	N	Y	Y	Y	Y	Y	Y
3.3 Major support	N	N	N	N	N	N	N	N	N	N	N	N	N
3.4 Suspension	N	N	N	N	N	N	N	N	N	N	N	N	N
	14	15	16	17	18	19	20	21	22	23	24	25	
1.1 Full	Y	N	N	Y	N	N	Y	N	N	Y	N	N	
1.2 Minor change	N	Y	N	N	Y	N	N	Y	N	N	Y	N	
1.3 Major change	N	N	Y	N	N	Y	N	N	Y	N	N	Y	
2. Legal action	N	N	N	Y	Y	Y	N	N	N	Y	Y	Y	
3.1 Full support	N	N	N	N	N	N	N	N	N	N	N	N	
3.2 Minor support	N	N	N	N	N	N	N	N	N	N	N	N	
3.3 Major support	Y	Y	Y	Y	Y	Y	N	N	N	N	N	N	
3.4 Suspension	N	N	N	N	N	N	Y	Y	Y	Y	Y	Y	

Source. Adapted from [7]

These alternatives (states for the conflict) are derived from positions *ex ante*, i.e. in a negotiation context, implementation prior to resolving the conflict. Due to the fact that the limitations eventually established by the authority ought to prevail and the desire of the proposers is to develop the project at the highest level allowed by the RLG, combinations with different levels of implementation allowed and proposed need not to be taken into account. In the same way, legal action by the Opposition need not be combined with the suspension of the project by the authority. Thus, from the 25 alternatives in Table 1, only alternatives 2, 5, 9, 12, 16, 19 and 20 need to be taken into account for the strategic analysis.

Indeed, Alternative 1 is a representation of the preliminary state of facts, without a position taken by the RLG. It is equivalent to allowing the full development of the project as contemplated by Alternative 2. Alternatives 3, 4, 6, 7, 8, 10, 11, 13, 14, 15, 17 and 18 need not be considered here because they present different levels of the proposal of the consortium and of support by the RLG. Nor need Alternatives 23, 24 and 25 be considered, because they combine suspension of the project by the authority with legal action by the opposition, legal action that becomes irrelevant if the initiative is not authorized. Therefore, the evaluations of the seven alternatives according to the four criteria, resulting from the assessments by the experts are given in Table 2. As above mentioned, applying the first two criteria the experts measure the benefits of the initiative being carried forward. On the other hand, the environmental and social impacts are considered on a basis of "the lower the better" by the RLG as well as by the NRP opposition.

Table 2. Evaluation matrix according to the four criteria

Alternative	Economic	Environmental	Social	Political
2	8	4	5	7
5	8	4	5	6
9	6	3	2	6
12	6	3	4	5
16	2	2	2	5
19	2	2	2	4
20	0	1	1	3

Source. Adapted from [7].

Given its position and power of authority, RLG takes into account the four criteria, while two criteria are relevant for the Consortium and the Opposition. Therefore, the criteria that are considered by the consortium are those which have evaluations that increase with the level of support by the RLG, i.e. the level of the economic and political impacts. In fact, the economic impact measures the size of the project effectively implemented, coincident with the level of implementation decided by the consortium in Table 1. The inclusion of the political impact is considered because this is the only factor whose value in Table 2 is consistently reduced if legal action is taken by the opposition. Finally, those criteria which measure negative environmental and social impacts are chosen by the NRP Opposition.

Thus, in the second stage, for the RLG, the composition was by the joint probability of not minimizing the positive economic and political impacts and not maximizing the negative environmental and social impacts. For the NRP Consortium, by the probability of not minimizing the number of jobs created and the political impacts. For the NRP opposition, by the probability of not maximizing the harmful environmental and social impacts.

Since the evaluations result from the contribution of six experts, beta distributions are used. The probabilities of each alternative being the most preferred and the least preferred according to each criterion assuming beta distributions with means in the observed evaluations in Table 2 are presented in Tables 3 and 4.

Table 3. Probabilities of each alternative being the most preferred according to each criterion

Alternative	Economic	Environmental	Social	Political
2	0.3967	0.3609	0.4346	0.452
5	0.3967	0.3609	0.4346	0.1857
9	0.0997	0.1112	0.0047	0.1857
12	0.0997	0.1112	0.116	0.0718
16	0.0035	0.0260	0.0047	0.0718
19	0.0035	0.0260	0.0047	0.0252
20	0.0003	0.0038	0.0005	0.0077

Table 4. Probabilities of each alternative being the least preferred according to each criterion

Alternative	Economic	Environmental	Social	Political
2	0.0012	0.0077	0.0008	0.0119
5	0.0012	0.0077	0.0008	0.0400
9	0.0099	0.0413	0.1521	0.0400
12	0.0099	0.0413	0.0067	0.1047
16	0.1852	0.1662	0.1521	0.1047
19	0.1852	0.1662	0.1521	0.2333
20	0.6074	0.5697	0.5356	0.4654

Table 5 presents the results of the application of CPP in the second stage, using the Pessimist-Conservative (PC) approach, to choose one alternative to represent each DM in the next stage. For instance, the score 0.3566 of alternative 2 for RLG is the product of four factors, the probabilities of not maximizing the preference according to the environmental and social criteria, 1-0.3609 and 1- 0.4346, respectively, and the probabilities of not minimizing the preference according to the economic and political criteria, 1- 0.0012 and 1-0.0119, respectively.

Table 5. Initial scores according to the three DMs

Alternative	RLG	Consortium	Opposition
2	0.3566	0.9869	0.3613
5	0.3464	0.9588	0.3613
9	0.8409	0.9506	0.8846
12	0.6966	0.8865	0.7857
16	0.7072	0.7295	0.9694
19	0.6055	0.6247	0.9694
20	0.2090	0.2099	0.9957

Therefore, the RLG most prefers Alternative 9, while the Consortium most prefers Alternative 2 and the Opposition most prefers Alternative 20. These three alternatives are then compared. The probabilities of maximizing the positive economic and political impacts and of minimizing the negative environmental and social impacts and, in the last column, the joint probabilities of maximizing the preference according to at least one of the four criteria, are presented in Table 6. For instance, the global score 0.9155 for alternative 2 is $1 - (1 - 0.7425) * (1 - 0.0277) * (1 - 0.0030) * (1 - 0.6616)$. As it can be seen, alternative 20 is the recommended one.

Table 6. Probabilistic evaluations and final scores for the three final alternatives

Alternative	Economic	Environmental	Social	Political	Global
2	0.7425	0.0277	0.0030	0.6616	0.9155
9	0.2545	0.1051	0.2545	0.3148	0.6592
20	0.0030	0.8672	0.7425	0.0236	0.9667

It is noticeable, nevertheless, that, according to the experts assessments in Table 1, Alternatives 2 and 9 are not very different from each other. Then in Stage 3, the Consortium might have agreed to the minor reduction of the project and withdrawn Alternative 2 to increase the chance of Alternative 9 in the final application of CPP. If that would have happened, this last alternative would have been chosen in the final stage.

5 Conclusion and Final Remarks

The application to the example showed that this methodology provides strategic ways to reduce the initial set of alternatives. This is especially important because the difficulties and effort involved in having to deal with a large number of alternatives is a known problem in complex conflict analysis. For instance, the authors of [10] point out that entering each DM's relative preferences over all feasible states is one of the most exacting challenges in the modelling stage of the GMCR.

Moreover, a clear advantage of this new methodology is that it is able to provide unique solutions for a conflict, as observed in the example. This result must be highlighted because most conflict analysis models usually generate more than one possible result for the conflict. Therefore, these models usually require a negotiation process after the conflict analysis or resolution process is ended. The CRMCPP methodology involves both the conflict analysis and the negotiation processes throughout its four stages.

Another important point is that applying CPP at different stages of negotiation to conflict resolution provides an objective basis for negotiators to move forward in their positions. Besides, it facilitates the acceptance of a final solution that is built quickly on solid bases. The modelling of the problem in terms of the composition of quantitatively evaluated criteria and the composition of preferences in terms of the probability of preference facilitates being better able not only to understand the reasons for differences but also being better able to identify proximities between different positions.

The main limitation of the new model is that the quality of the results will depend decisively on the efficiency in the initial selection of the criteria and the evaluation of the alternatives according to the different criteria. However, the transformation of the initial data into probabilities of reaching the extremes of best and worst evaluation reduces the influence of errors of evaluation of less preferred alternatives and facilitates the interpretation of the data. For future research, it is suggested that the results of this new methodology be compared with equilibrium concepts of the GMCR.

References

1. Von Neumann, J., Morgenstern, O.: Theory of Games and Economic Behavior. Princeton University Press, Princeton (1944)
2. Howard, N.: Paradoxes of Rationality: Theory of Metagames and Political Behavior. MIT Press, Cambridge (1971)
3. Fraser, N.M., Hipel. K.W.: Conflict Analysis: Models and Resolutions. North-Holland, New York (1984)
4. Kilgour, D.M., Hipel, K.W., Fang, L.: The graph model for conflicts. Automatica **23**(1), 41–55 (1987)
5. Xu, H., Hipel, K.W., Kilgour, D.M., Fang, L.: Conflict Resolution Using the Graph Model: Strategic Interactions in Competition and Cooperation. SSDC, vol. 153. Springer, Cham (2018). https://doi.org/10.1007/978-3-319-77670-5
6. Sant'Anna, A.P.: Probabilistic Composition of Preferences, Theory and Applications. DE. Springer, Cham (2015). https://doi.org/10.1007/978-3-319-11277-0
7. Silva, M.M., Kilgour, D.M., Hipel, K.W., Costa, A.P.C.S.: Probabilistic composition of preferences in the graph model with application to the new recife project. J. Legal Aff. Dispute Resol. Eng. Constr. **9**(3), 1–13 (2017)
8. Almeida, A.T., Morais, D.C., Daher, S.F.D.: Group Decision and Negotiation. Editora Universitária, Recife (2012)
9. Raiffa, H., Richardson, J., Matcalfe, D.: Negotiation Analysis: The Science and Art of Collaborative Decision Making. Harvard University Press, Cambridge (2002)
10. Yu, J., Hipel, K.W., Kilgour, D.M., Zhao, M.: Option prioritization for unknown preference. J. Syst. Sci. Syst. Eng. **25**(1), 39–61 (2016)

Analysis of Disputed Territories
in the Barents Sea

Sergey Demin[1,2]($\boxtimes$) (ID) and Sergey Shvydun[1,2] (ID)

[1] National Research University Higher School of Economics,
Myasnitskaya Str. 20, 101000 Moscow, Russia
sdemin@hse.ru
[2] V.A. Trapeznikov Institute of Control Sciences of Russian Academy
of Science, Profsoyuznaya Str. 65, 117342 Moscow, Russia

Abstract. As a result of the global warming, the situation in the Barents Sea leads to several important consequences. Firstly, oil and gas drilling becomes much easier than before. Therefore, it may raise the level of discussions on disputed shelf zones where the deposits are located, especially near to Norway-Russia sea border. Secondly, oil and gas excavation leads to potential threats to fishing by changing natural habitats, which in turn can create serious damage to the economies.

We construct a model, which helps to highlight potential disputed territories and analyze preferences of the countries interested in fossil fuels and fish resources. We also compare different scenarios of resource allocation with allocation by current agreement.

Keywords: The Barents Sea · Oil and gas deposits · Fishing resources ·
Disputed territories

1 Introduction

Over the past 20–30 years, the share of oil and gas in the global fuel and energy balance of consumption is more than 70% of all types of energy sources. Exploration for oil and gas is produced on the shelf in more than 70 countries. Meanwhile, global warming has made the territories covered with snow and glaciers more accessible for resource exploitation, thus, resulting in the increased interest in these areas.

Certainly, there are some regulations which help to divide territories beyond continental coasts. Firstly, territorial waters – belt of sea or ocean not exceeding 12 nautical miles, measured from the coast, adjacent to the coast under the sovereignty of the coastal state or its internal waters – give for this state the sovereignty beyond the territory [1].

Secondly, it takes into account exclusive economic zones (EEZ), which were formally introduced in 1982 by the UN Convention on the Law of the Sea [1]. EEZ is an area beyond the territorial sea, extending seaward to a distance of no more than 200 nautical miles (370 km) from its coastal baseline. This territory can be used by the other countries only for transportation. Meanwhile, a coastal state has sovereign rights for the purpose of exploration, development, conservation and management of natural

© Springer Nature Switzerland AG 2020
D. C. Morais et al. (Eds.): GDN 2020, LNBIP 388, pp. 32–44, 2020.
https://doi.org/10.1007/978-3-030-48641-9_3

resources, both living and nonliving, in the waters covering the seabed and other activities for economic exploration and exploitation of the zone, such as production of energy from water, currents and winds. The exception to this rule occurs when the exclusive economic zone will be crossed. When an overlap occurs, it is up to the states to specify the actual maritime boundary.

In this paper, we are focused on the Barents Sea which pertains to Russia and Norway. The area in the central part of the Barents Sea, called the Loop Hole, is the area beyond and totally enclosed from the 200 nautical miles limits of Norway and the Russian Federation [2].

As a result, they disputed EEZ in the Barents Sea. The treaty was agreed only in April 2010 between the two states and subsequently (September 15th, 2010) ratified, resolving this dispute [3]. The maritime delimitation was defined by the group of geodesic lines between certain points defined by the following coordinates:

1) 70° 16' 28.95" N, 32° 04' 23.00" E
2) 73° 41' 10.85" N, 37° 00' 00.00" E
3) 75° 11' 41.00" N, 37° 00' 00.00" E
4) 75° 48' 00.74" N, 38° 00' 00.00" E
5) 78° 37' 29.50" N, 38° 00' 00.00" E
6) 79° 17' 04.77" N, 34° 59' 56.00" E
7) 83° 21' 07.00" N, 35° 00' 00.29" E
8) 84° 41' 40.67" N, 32° 03' 51.36" E

This solution (see Fig. 1) established conditions for fishing cooperation, providing for the retention of the mechanism to jointly regulate fishing in the Barents Sea. In addition, the principles of cooperation in fossil fuels deposits exploration were also defined [3].

Fig. 1. Norway-Russia sea border.

Although the problem of disputable zones in the Barents Sea have already been solved, negotiations took 36 years [4]. As there are many other territorial disputes, we consider the problem as a fair division one. Our goal is to analyze territorial dispute in the Barents Sea and construct various scenarios of areas allocation taking into account different types of resources. The problem lies in the evaluation of the utility of each area for each agent as well as the influence of a disputed territory in order to find some allocation that satisfies all interested parties.

In Sect. 2 data used in our survey are described. The main idea of the constructed model for territory division is presented in Sect. 3. Section 4 explains main scenarios analyzed in the work. In Sect. 5 all results of the model application are demonstrated. Finally, Sect. 6 concludes the work.

2 Data Description

For application of our model we required the data considering location of all resources which can be interesting for the countries in this region. Each area is located at some distance from each country and possesses some natural resources. We consider two main natural resources according to [5]: fossil fuels such as oil and gas (O&G) and fish (F) resources. It is necessary to note that since these resources require maritime access to the areas, we did not consider shipping routes as one more additional resource.

For the data considering the location of fish resources we used an interactive map of the Barents Sea [6], wherein we can highlight fishing territories for different periods (for instance, see Fig. 2).

Fig. 2. Average fishing territories in the first quarter of the year.

In turn, as for the oil and gas data, we used information from [7]. In this source we have the following map (see Fig. 3) displaying territories with potential for finding oil and natural gas and confirmed oil and gas fields.

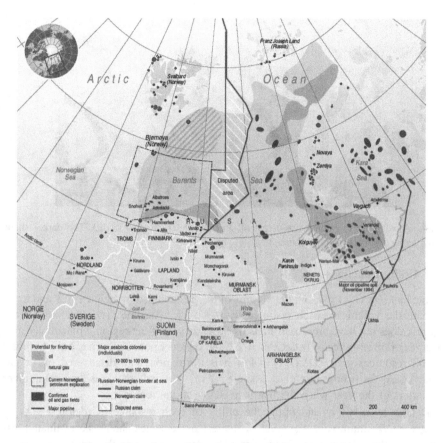

Fig. 3. Map of potential and confirmed oil and gas fields.

3 A Model

3.1 Problem Statement

Consider a set of areas X in the Barents Sea characterized by a set of parameters K and a set of countries Y which are interested in these areas.

Since the level of interest in each area of the sea is different, let us divide the whole region into some sub-regions almost of equal part. In our paper the Barents Sea was divided into 841,995 equal areas where each area encompasses a territory of about 210,000 square meters. Among them, only 241,162 areas have oil, gas or fish deposits.

Based on recent studies on natural resources availability in the Barents Sea [7] we can demonstrate that information in Fig. 4.

Let us evaluate the level of interest of each country in the Barents Sea.

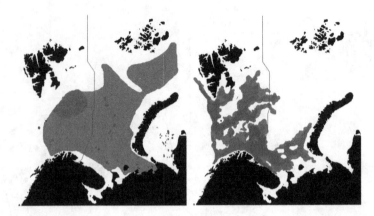

Fig. 4. Availability of natural resources: oil and gas (left) and fish (right). The darker shade means larger reserves of the resource.

3.2   Utility Functions

Denote by $f(O\&G, x), f(F, x)$, the volume of oil, gas and fish in region $x \in X$ and let us estimate the volume of each resource by 0–4 scale. Assume $f(O\&G, x) = 0$ if the region x does not have any fossil fuels, $f(O\&G, x) = 1$ if the region x potentially may have gas or oil resources, $f(O\&G, x) = 2$ if the region x potentially may have both gas and oil resources and $f(O\&G, x) = 4$ if it is the region with discovered oil or gas resources. As for fishing resources, assume $f(F, x) = m$, where m – is the total number of quarters of the year when the fish is available in the region x.

Denote by $u_k^{O\&G}(x)$ and $u_k^F(x)$ the utility of each resource in region $x \in X$ for country $k \in Y$. Intuitively, the level of interest of all zones should be evaluated differently for the same country. Moreover, among two areas with the same quantities of natural resources, the priority should be given to the closest one. Following [8, 9], we assume that the interest of a country in each resource is proportional to the distance to the area and equal to zero after some distance d^* (hereafter we use $d^* = 1000$ km). Then the interest of each country in natural resources located in some area is characterized by the following formulae

Fossil Fuels

$$u_k^{O\&G}(x) = \begin{cases} f(O\&G, x) \cdot \left(\frac{d^* - d_k(x)}{d^*}\right), & if \quad d_k(x) < d^*, \\ 0, & if \quad d_k(x) \geq d^* \end{cases} \tag{1}$$

Fish

$$u_k^F(x) = \begin{cases} f(F, x) \cdot \left(\frac{d^* - d_k(x)}{d^*}\right), & if \quad d_k(x) < d^*, \\ 0, & if \quad d_k(x) \geq d^*, \end{cases} \tag{2}$$

where $d_k(x)$ is the distance from the closest point of the country $k \in Y$ to the area $x \in X$. The total utility of each area $u_k^T(x)$ is calculated as

$$u_k^T(x) = \propto \cdot u_k^{O\&G}(x) + u_k^F(x). \tag{3}$$

Where $\propto$ is a coefficient that characterizes the relative importance of fossil fuels compared to fish resources. It should be mentioned here, that generally each country might evaluate natural resources differently, based on its industrial base, needs of the economy, etc. However, for the simplicity, we assume that each country evaluates each resource equally.

Thus, we can evaluate an interest of each country in a specific area of the Barents Sea and find regions of the most interest for each country. The areas can also be ranked lexicographically or by some other procedures.

3.3 Areas Allocation

According to treaty between the Kingdom of Norway and the Russian Federation concerning maritime delimitation and cooperation in the Barents Sea and the Arctic Ocean [3], the maritime borders in the Barents Sea are fixed now and, thus, there are no disputed areas in the region. Moreover, if we assume that current borders are the equilibrium for two states, i.e., each country is satisfied with present delimitation of areas, we can evaluate the relative importance of fossil fuels compared to fish resources.

Let us define the fairness of the allocation. The fairness of the allocation can be evaluated differently; in our paper it is based on the satisfaction level of each country $S_k(P)$ which is calculated as [9]

$$S_k(P) = \sum_{x \in X:(x,k) \in P} \left(u_k^T(x) \right) - \sum_{x \in X:(x,k) \notin P} \left(u_k^T(x) \right), \tag{4}$$

where P is a binary relation $P \subset X \times Y$ that characterizes the final allocation of areas. In other words, the satisfaction level of a country is calculated as the difference between the total utility of areas that were allocated to this country, and the potential total utility of areas that were not allocated to the country.

If we assume that actual allocation is fair for Norway and Russia and both countries have the same interest in natural resources, then, according to our model, the coefficient $\propto$ is equal to 1.74. In other words, the importance of fossil fuels for countries is 1.74 times higher than the importance of fishing resources.

Since natural resources that we consider are limited, the availability of resources may change over time. Moreover, there may be some changes in global energy markets which means that the relative importance of natural resources may differ resulting in a potential disputed territory in the Barents Sea. Thus, it will be valuable to consider different scenarios of how the relative importance of natural resources may change in order to evaluate the sustainability of the present allocation of zones.

4 Resolution Models

Next, we propose several models of areas allocation, which are fair in some sense for each country, evaluate the satisfaction level of each country and consider different valuations of parameter $\propto$.

4.1 Allocation of Areas Regardless the Level of Interest in Areas of the Barents Sea

Scenario 1: All Areas are Allocated with Respect to the Current Borders
Since the borders in the Barents Sea are clearly defined, let us consider the allocation with respect to the borders. The results are provided in Fig. 5.

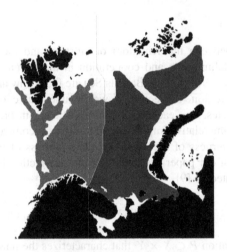

Fig. 5. Allocation according to scenario 1 (areas allocated to Norway are colored in red (left part), to Russia – in blue (right part)). (Color figure online)

Scenario 2: All Areas are Allocated with Respect to the Distance.
Let us allocate all areas to the country which closer located to it. The results are provided in Fig. 6.

4.2 Allocation of Areas with Respect to the Level of Interest in Areas of the Barents Sea

Next, let us consider two models of areas allocation which are based on the level of interest of each country. According to the first model, the allocation of areas is performed similarly to the adjusted winner procedure [10]. The second procedure allocates areas to the most interested party.

Fig. 6. Allocation according to scenario 1 (areas allocated to Norway are colored in red (left part), to Russia – in blue (right part)). (Color figure online)

Modified Adjusted Winner Procedure

Since in our case there are only two countries interested in the region, we can implement the adjusted winner procedure which is used in fair division problems for the case of two agents. The adjusted winner procedure ensures that the final allocation is proportional (each side receives a piece that he/she perceives to be at least 1/n of the whole), envy free (no agent has incentives to exchange his allocated part of the object with any other agent) and Pareto-optimal (no other allocation can make one party better off without making the other party worse off).

Generally, the adjusted winner procedure is used for fair division of divisible goods. However, since we do not allow shared allocation of areas, the final allocation of areas does not guarantee that the satisfaction level of each country will be equal. Thus, the final allocation of zones is not unique, and we need to consider different initial allocations of zones.

Scenarios 3, 4: Initial Allocation to Norway and Russia (Correspondingly)

The model of disputed territory resolution works as follows. Suppose we have some initial allocation of areas. Then we can evaluate the satisfaction level of each country. If the satisfaction level is equal, the procedure of areas allocation stops. Otherwise, the exchange procedure is performed between two countries. Denote by k_1 and k_2 the most unsatisfied and the most satisfied countries. Then the exchange procedure is performed for the area $x \in X$ which satisfies the following conditions:

$$(x, k_2) \in P, \tag{5}$$

$$u_{k_1}^T(x) \neq 0, \tag{6}$$

$$\frac{u_{k_1}^T(x)}{u_{k_2}^T(x)} \to max. \tag{7}$$

The criterion for the choice of exchanging area $x \in X$ between countries is similar to the criterion used for adjusted winner procedure [10]. First, the area $x \in X$ should belong to the most satisfied country. Second, the area $x \in X$ should be valuable for unsatisfied country. Finally, the exchange should be performed for the area which is valuable as much as possible for unsatisfied country and as less as possible for satisfied country.

Thus, some new allocation is obtained and the whole procedure repeats again. There are different criteria that can be used to terminate the exchange procedure. In our paper the procedure stops if there are no areas available for the exchange procedure or the most unsatisfied country is changing at each new step of the exchange procedure.

4.3   Allocation of Areas to the Most Interested Country

Scenario 5: All Areas are Allocated to the Country That Values Them the Most
The allocation of zones is performed by a simple majority rule [11]. In other words, a disputed territory is allocated to country B if the total number of resources which country B is interested in more than country A is more than or equal to 50% + 1 of the total number of resources available in this zone.

5   Results

Now let us apply each model and compare the results.

The results for scenarios 1–2 are provided in Fig. 5 and 6. As for other scenarios, since we consider two different resources (fossil fuels and fish resources) which are not always equally valued by countries, we should consider the following cases

5.1   Fossil Fuels and Fish Resources Have the Same Importance ($\alpha = 1$)

The results of the models are provided in Fig. 7.

Fig. 7. Allocation according to scenarios 3–5 for $\alpha = 1$ (areas allocated to Norway are colored in red (left part), to Russia – in blue (right part)). (Color figure online)

Now let us evaluate the efficiency of each model in terms of the total satisfaction level. The satisfaction level of each country according to different scenarios is provided in Table 1.

Table 1. Satisfaction level.

№	Norway	Russia	Total
Scenario 1	100625	46833	147458
Scenario 2	126655	23833	150488
Scenario 3	74461	74446	148907
Scenario 4	74451	74455	148906
Scenario 5	126655	23833	150488

According to Table 1, if gas and oil have the same importance as fish, Norway is the most satisfied country according to the current allocation of areas (scenario 1) or if the areas will be allocated with respect to the distance (scenario 2) or to the most interested party (scenario 5). As for the total satisfaction level, the allocation with respect to the distance or to the most interested party showed the highest values than any other scenario while current allocation of zones is the worst one. However, if we choose the difference in satisfaction level as criterion of the efficiency of the model, scenarios 3-4 are the best allocations since both countries have almost the same satisfaction level. Finally, we can see that scenarios 2 and 5 result in the same allocation of areas. It can be explained by the fact that the level of interest is evaluated with respect to the distance.

5.2   Fossil Fuels Are Five Times More Important Than Fish Resources ($\alpha = 5$)

The results of the models are almost the same and provided in Fig. 8.

Fig. 8. Allocation according to scenarios 3–5 for $\alpha = 5$ (areas allocated to Norway are colored in red (left part), to Russia – in blue (right part)). (Color figure online)

Now let us evaluate the efficiency of each model in terms of the total satisfaction level. The satisfaction level of each country according to different scenarios is provided in Table 2.

Table 2. Satisfaction level.

№	Norway	Russia	Total
Scenario 1	102877	341267	444144
Scenario 2	210150	244724	454874
Scenario 3	227406	227381	454787
Scenario 4	227383	227404	454787
Scenario 5	210150	244724	454874

According to Table 2, if gas and oil is 5 times more important than fish, Russia is the most satisfied country according to the current allocation of areas (scenario 1) or if the areas will be allocated with respect to the distance (scenario 2) or to the most interested party (scenario 5). As for the total satisfaction level, scenarios 2–5 have almost the same values while current allocation of zones is the worst one. However, if we choose the difference in satisfaction level as criterion of the efficiency of the model, scenarios 3-4 are the best allocations since both countries have almost the same satisfaction level.

5.3   Fossil Fuels Are Ten Times More Important Than Fish Resources ($\alpha = 10$)

The results of the models are provided in Fig. 9.

Fig. 9.   Allocation according to scenarios 3–5 for $\alpha = 10$ (areas allocated to Norway are colored in red (left part), to Russia – in blue (right part)). (Color figure online)

Now let us evaluate the efficiency of each model in terms of the total satisfaction level. The satisfaction level of each country according to different scenarios is provided in Table 3.

Table 3. Satisfaction level.

№	Norway	Russia	Total
Scenario 1	105691	709309	815001
Scenario 2	314519	520839	835358
Scenario 3	416719	416713	833433
Scenario 4	416703	416730	833433
Scenario 5	314519	520839	835358

According to Table 3, if gas and oil is 10 times more important than fish, Russia is the most satisfied country according to the current allocation of areas (scenario 1) or if the areas will be allocated with respect to the distance (scenario 2) or to the most interested party (scenario 5). As for the total satisfaction level, scenarios 2–5 have almost the same values while current allocation of zones is the worst one. However, if we choose the difference in satisfaction level as criterion of the efficiency of the model, scenarios 3-4 are the best allocations since both countries have almost the same satisfaction level.

6 Conclusion

We considered the problem of potentially disputed territories resolution in the Barents Sea. Using an introduced model of utility values with respect to main resources – fossil fuels and fish – we have proposed different scenarios for allocation of territories depending on the importance of each natural resource. As a result, we have evaluated the satisfaction level of countries according to each scenario and proposed allocation with the same satisfaction level of each country and compared it with current allocation. We hope that application of fair division models will help in resolving conflicts in other parts of the world.

Acknowledgements. The paper was prepared within the framework of the HSE University Basic Research Program and funded by the Russian Academic Excellence Project '5-100'. The authors also thank Professor Fuad Aleskerov and the International Center of Decision Choice and Analysis (DeCAn Center) of the National Research University Higher School of Economics for their support of the work.

References

1. UN General Assembly: Convention on the Law of the Sea, 10 December 1982. http://www.refworld.org/docid/3dd8fd1b4.html/. Accessed 18 Mar 2020
2. Subcommission established for the consideration of the submission made by Norway: Summary of the Recommendations of the Commission on the Limits of the Continental Shelf in Regard to the Submission Made by Norway in Respect of Areas in the Arctic Ocean, the Barents Sea and the Norwegian Sea, 27 November 2006 (2006)
3. Treaty between the Kingdom of Norway and the Russian Federation concerning Maritime Delimitation and Cooperation in the Barents Sea and the Arctic Ocean (2010)
4. Eu-arctic-forum.org: Delimitation Agreement: A New Era In The Barents Sea And The Arctic?—Arctic Forum Foundation (2020). http://eu-arctic-forum.org/allgemein/delimitation-agreement-a-new-era-in-the-barents-sea-and-the-arctic/. Accessed 18 Mar 2020
5. WWF Russia: About The Barents Region (2020). https://wwf.ru/en/regions/the-barents-branch/o-barents-regione/. Accessed 18 Mar 2020
6. Projects.scanex.ru: Scanex Web Geomixer - Просмотр Карты (2017). http://projects.scanex.ru/RussianArcticMSP-Barents/. Accessed 18 Mar 2020
7. Rekacewicz, P.: UNEP/GRID - Arendal. http://www.grida.no/resources/7482/. Accessed 18 Mar 2020
8. Aleskerov, F., Victorova, E.: An analysis of potential conflict zones in the Arctic Region. Working paper WP7/2015/05. Moscow: HSE Publishing House (2015). 20 p
9. Aleskerov, F., Shvydun, S.: Group Decis. Negot. **28**, 11 (2019). https://doi.org/10.1007/s10726-018-9596-4
10. Brams S.J., Taylor A.D.: Fair Division: From Cake-Cutting to Dispute Resolution. Cambridge University Press, Cambridge (1996)
11. Aleskerov, F.T., Subochev, A.: Modeling optimal social choice: matrix-vector representation of various solution concepts based on majority rule. J. Glob. Optim. **56**(2), 737–756 (2013)

A Novel Method for Eliminating Redundant Option Statements in the Graph Model for Conflict Resolution

Shinan Zhao[1](✉) 🆔 and Haiyan Xu[2]

[1] School of Economics and Management,
Jiangsu University of Science and Technology, Zhenjiang 212003, China
zhugeliuyun1989@163.com
[2] College of Economics and Management,
Nanjing University of Aeronautics and Astronautics, Nanjing 211106, China

Abstract. The option prioritization is the most effective preference ranking approach within the framework of the graph model for conflict resolution, in which a set of option statements for each decision maker (DM) involved in a dispute is determined by individual judgments. Inevitably, some option statements may be unnecessary or redundant. To address the redundancy of option statements, a novel option statement reduction method as well as an effective reduction algorithm is developed in this research based on the rough set theory. Furthermore, the Elmira conflict is utilized to show how the proposed option statement reduction method can be employed for efficiently eliminating redundant option statements of DMs.

Keywords: Graph model for conflict resolution · Option statements · Redundancy · Preference ranking

1 Introduction

Conflicts are very pervasive in social, political, economic, environmental and other areas, where multiple stakeholders or decision makers (DMs) involved in a given dispute dynamically interact with each other for pursuing their own benefits. The graph model for conflict resolution (GMCR) is a very powerful and flexible methodology which can be employed for modeling and analyzing tough conflict situations [1–3]. In the modeling stage, the DMs, their options, feasible states and each DM's preference over states should be identified according to the background information about a particular conflict. The preference of DMs is very important in stability analysis but difficult to be determined due to the diversity of individual value systems. Three preference ranking methods were developed by Fang et al. [4, 5] within the framework of GMCR for conveniently acquiring DMs' preference over states: direct ranking, option weighting and option prioritization.

The most commonly used and effective technique for preference ranking in GMCR is option prioritization, in which each DM's preference over states can be reflected by a set of option statements consisting of some numbered options and connectives. Then

© Springer Nature Switzerland AG 2020
D. C. Morais et al. (Eds.): GDN 2020, LNBIP 388, pp. 45–55, 2020.
https://doi.org/10.1007/978-3-030-48641-9_4

the classical option prioritization approach was improved for handing strength of preference [6–8], unknown preference [9, 10], fuzzy preference [11–14], grey preference [15, 16], and probabilistic preference [17, 18]. Since the option statements are determined according to individual cognition and subjectivity, there may exist some unnecessary or redundant statements which should be removed for the sake of computing efficiency. Hence, a novel option statement reduction approach is proposed in this research for eliminating redundant option statements in option prioritization.

The organization of this paper is as follows. In Sect. 2, the option prioritization technique is briefly introduced within the GMCR paradigm. Subsequently, an option statement reduction method as well as its algorithm is purposefully developed in Sect. 3. In Sect. 4, the Elmira conflict is utilized for demonstrating the practicality of the proposed method. Finally, some conclusions and future work are presented in Sect. 5.

2 Option Prioritization in GMCR

A conflict can be modeled as a graph model $G = \langle N, S, \{A_i, \succsim_i : i \in N\} \rangle$, where

(1) $N = \{1, 2, \cdots, i, \cdots, n\}$ is the set of decision makers;
(2) $S = \{s_1, s_2, \cdots, s_l, \cdots, s_m\}$ is the set of feasible states;
(3) A_i is the set of oriented arcs of DM $i \in N$, which contains all of the unilateral moves by DM i in one step; and
(4) $\succsim_i$ stands for the preference relations (more or less preferred) over states by DM i.

Within the framework of GMCR, the option prioritization is the most convenient and effective technique for acquiring the preference of DMs involved in a dispute, in which a DM's preference over states can be reflected by a set of option statements composed of numbered options and several logical connectives such as "& (and)", "- (not)", "| (or)", "IF" and "IFF (if and only if)". Furthermore, the option statements are presented from the most to least important in a hierarchical order.

Let $K = \{\Omega^1, \Omega^2, \cdots, \Omega^l, \cdots, \Omega^k\}$ be the set of option statements listed by priority for a given DM, in which Ω^l is the lth option statement. In a particular state $s \in S$, each option statement Ω^l can be true (T) or false (F). Let the value of Ω^l be $\Omega^l(s)$. If Ω^l holds at state s, then $\Omega^l(s) = T$; otherwise, $\Omega^l(s) = F$.

The incremental score of the option statement Ω^l at state s is written as

$$\Psi_l(s) = \begin{cases} 2^{k-l} & \text{if } \Omega_l(s) = T \\ 0 & \text{otherwise} \end{cases} \tag{1}$$

The total score of all of the option statements in K at state s is denoted by

$$\Psi(s) = \sum_{l=1}^{k} \Psi_l(s) \tag{2}$$

According to each state's score $\Psi(s)$, a DM's preference over states can be determined. More specifically, a state with a higher score is more preferred than a state with a lower score. And two states are equally preferred if their scores are the same.

As introduced above, a DM's preference over states can be easily obtained by using the option prioritization technique. However, some option statements may be redundant since all of the option statements are determined according to personal judgments. The redundancy of option statements which should be removed could increase the computational complexity of preference ranking, especially in large-scale conflicts. Hence, it is very important to develop an option statement reduction method for determining minimal option statement sets that do not change the preference ranking results. In the following, an attribute reduction method based on rough set theory [19, 20] is incorporated into the option prioritization within the GMCR paradigm for eliminating redundant or useless option statements.

3   An Option Statement Reduction Method for Option Prioritization

The objective of the option statement reduction method is to find minimal option statements sets which do not change the results of preference ranking. The basic idea is as follows:

(1)   Determine the universe, condition and decision attributes

The feasible states, option statements, and the orders of states are regarded as universe (U), condition attributes (C) and decision attribute (D), respectively. Let $S = \{s_1, s_2, \cdots, s_m\}$ be the set of feasible states, $K = \{\Omega^1, \Omega^2, \cdots, \Omega^l, \cdots, \Omega^k\}$ be the set of option statements listed from the most important on the left to the least important on the right for a given DM, $D_0 = \{d_1, d_2, \cdots, d_m\}$ be the order of ranked states, and a_{ij} be the truth value (T or F) of the option statement Ω^j at state s_i. The information system for option statements can be represented as shown in Table 1.

Table 1. The information system for option statements

Universe (U)	Condition attributes (C)				Decision attributes (D)
	Ω^1	Ω^2	...	Ω^k	d_i (the order of ranked states)
s_1	a_{11}	a_{12}	...	a_{1k}	d_1
s_2	a_{21}	a_{22}	...	a_{2k}	d_2
...	...	...	...	...	...
s_m	a_{m1}	a_{m2}	...	a_{mk}	d_m

(2)   Calculate the option statement reducts based on rough set theory

According to Table 1, the option statement reducts can be determined by using attribute reduction methods in rough set theory. Furthermore, ROSETTA, a toolkit for analyzing minimal attribute sets within the framework of rough set theory, can be conveniently utilized to obtain the minimal option statement sets. Note that not all option statement reducts are the satisfactory solutions because the order of ranked states by using a statement reduct may be not the same as the preference ranking produced by the initial set of option statements.

(3) **Filter the option statement reducts that can generate the same order of ranked states**

One can calculate the order of ranked states for each option statement reduct to check whether the order is equal to the initial one. And only the option statement reducts that do not change the preference ranking results are kept.

Let D_0 be the order of ranked states by using the initial set of option statements for a given DM, and $A_1, A_2, …, A_i$ be the option statement reducts. The procedure for eliminating redundant option statements in option prioritization can be summarized in Fig. 1.

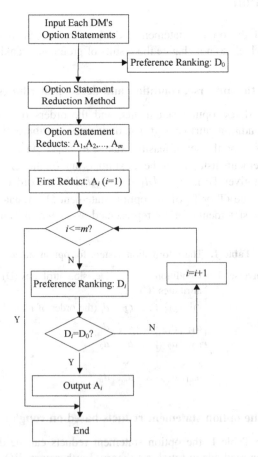

Fig. 1. The algorithm of option statement reduction method

4   Case Studies

A real-world groundwater contamination dispute occurred in Elmira, a small agricultural town in Southern Ontario, Canada, is investigated in this section to show how the proposed option statement reduction method can be applied in reality for eliminating unnecessary or redundant option statements due to individual subjectivity.

Underlying aquifers provided Elmira with safe and fresh drinking water. In 1989, the N-Nitrosodimethylamine (NDMA), a chemical substance which could cause cancer, was discovered in the Elmira's aquifer supplying water by the Ontario Ministry of the Environment (MoE). Uniroyal Chemical Ltd. (UR) was suspected as being the polluter who generated the contamination found in the aquifers. UR was requested to take effective measures for cleaning up the pollutants according to the Ontario's environmental laws (Control Order) issued by MoE [3]. In the Elmira conflict, there are three DMs and five options [21, 22]:

- Ontario Ministry of the Environment (**MoE**) has one option that whether or not to **modify** the original Control Order.
- Uniroyal Company (**UR**) has three options: (1) **Delay** the negotiations by "dragging its feet"; (2) **Accept** the current Control Order; (3) **Abandon** its Elmira plant.
- Local Government (**LG**) has one option that whether or not to **insist** on the initial Control Order.

After removing infeasible states, only nine feasible states are left as shown in Table 2.

Table 2. DMs, options and feasible states in the Elmira conflict.

DMs	Options	Feasible states								
MoE	1. Modify	N	Y	N	Y	N	Y	N	Y	-
UR	2. Delay	Y	Y	N	N	Y	Y	N	N	-
	3. Accept	N	N	Y	Y	N	N	Y	Y	-
	4. Abandon	N	N	N	N	N	N	N	N	Y
LG	5. Insist	N	N	N	N	Y	Y	Y	Y	-
State number		s_1	s_2	s_3	s_4	s_5	s_6	s_7	s_8	s_9

In the Elmira conflict, each DM's preference over states can be described by using a set of option statements as given in Tables 3, 4 and 5.

Table 3. Option statements and interpretations of MoE

Option statement	Interpretation
−4	MoE does not want UR to abandon its operation in Elmira
3	MoE wants UR to accept a control order
−2	MoE does not like to see the delay of UR
−1	MoE does not want to modify the original order
5 IFF−1	MoE wants LG to insist on the original order if and only if he chooses not to modify

Table 4. Option statements and interpretations of UR

Option statement	Interpretation
3 IFF1	UR will accept the control order if and only if MoE chooses to modify the original order
−4	UR does not want to abandon its operation in Elmira
−5	UR does not like that LG insist on the original order
2 IFF−5	UR would like to delay if and only if LG prefers not to insist on the original order

Table 5. Option statements and interpretations of LG

Option statement	Interpretation
−4	LG does not want UR to abandon its operation in Elmira
−1	LG prefers that the original control order not be modified
3 IF−1	LG wants UR to accept the original order if MoE does not modify it
5 IF1	LG would insist on the original order if MoE tends to modify it
−2	LG does not want UR to delay the procedure
5	LG wants to insist on the original control order

By using the option prioritization technique, each DM's preference over states can be determined as follows.

(1) MoE's preference is $s_7 \succ s_3 \succ s_4 \succ s_8 \succ s_5 \succ s_1 \succ s_2 \succ s_6 \succ s_9$.
(2) UR's preference is $s_1 \succ s_4 \succ s_8 \succ s_5 \succ s_9 \succ s_2 \succ s_3 \succ s_7 \succ s_6$.
(3) LG's preference is $s_7 \succ s_3 \succ s_5 \succ s_1 \succ s_8 \succ s_6 \succ s_4 \succ s_2 \succ s_9$.

As shown in Tables 3, 4 and 5, the number of option statements for MoE, UR and LG are 5, 4 and 6, respectively, in which some option statements may be redundant and should be eliminated. The option statement reduction method proposed in Sect. 3 can be employed for removing redundant option statements.

Firstly, we can calculate MoE's minimal option statement reduct sets. According to Table 3 and MoE's preference, MoE's information system of option statements is expressed as given in Table 6.

Table 6. MoE's information system of option statements

Universe (U)	Condition attributes (C)					Decision attributes (D)
	a1 = −4	a2 = 3	a3 = −2	a4 = −1	a5 = 5 IFF−1	d_i (the order of ranked states)
s_1	T	F	F	T	F	6
s_2	T	F	F	F	T	7
s_3	T	T	T	T	F	2
s_4	T	T	T	F	T	3
s_5	T	F	F	T	T	5
s_6	T	F	F	F	F	8
s_7	T	T	T	T	T	1
s_8	T	T	T	F	F	4
s_9	F	F	T	T	F	9

By using the ROSETTA software, the option statement reducts of MoE can be conveniently obtained as illustrated in Fig. 2.

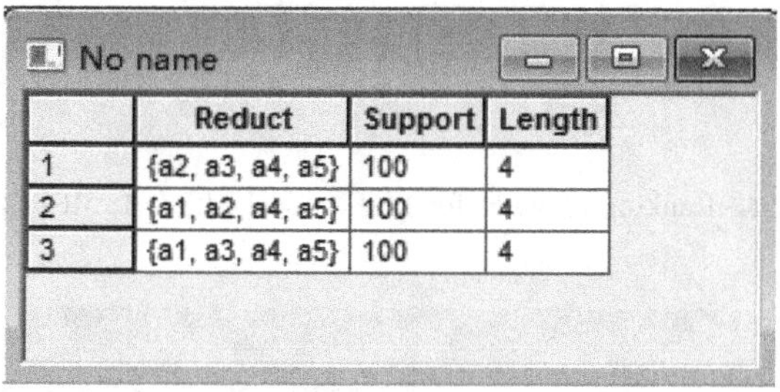

	Reduct	Support	Length
1	{a2, a3, a4, a5}	100	4
2	{a1, a2, a4, a5}	100	4
3	{a1, a3, a4, a5}	100	4

Fig. 2. Option statements reducts for MoE

According to Fig. 2, there are three reducts for MoE's option statements:

(1) A_1 = {a2, a3, a4, a5}, which means that the option statement set is {3, −2, −1, 5IFF−1}.
(2) A_2 = {a1, a2, a4, a5}, which means that the option statement set is {−4, 3, −1, 5IFF−1}.
(3) A_3 = {a1, a3, a4, a5}, which means that the option statement set is{−4, −2, −1, 5IFF−1}.

By using GMCR II software [4, 5], the ranking of states for MoE under the aforementioned three reducts can be determined as given in Figs. 3, 4 and 5.

DMs	Options	7	3	4	8	5	1	2	9	6
MOE	1. Modify	N	N	Y	Y	N	N	Y	—	Y
UR	2. Delay	N	N	N	N	Y	Y	Y	—	Y
	3. Accept	Y	Y	Y	Y	N	N	N	—	N
	4. Abandon	N	N	N	N	N	N	N	Y	N
LG	5. Insist	Y	N	N	Y	Y	N	N	—	Y

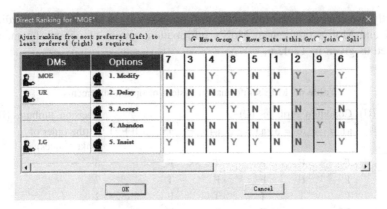

Fig. 3. Ranking of states for MoE under {3, −2, −1, 5IFF−1}

DMs	Options	7	3	4	8	5	1	2	6	9
MOE	1. Modify	N	N	Y	Y	N	N	Y	Y	—
UR	2. Delay	N	N	N	N	Y	Y	Y	Y	—
	3. Accept	Y	Y	Y	Y	N	N	N	N	—
	4. Abandon	N	N	N	N	N	N	N	N	Y
LG	5. Insist	Y	N	N	Y	Y	N	N	Y	—

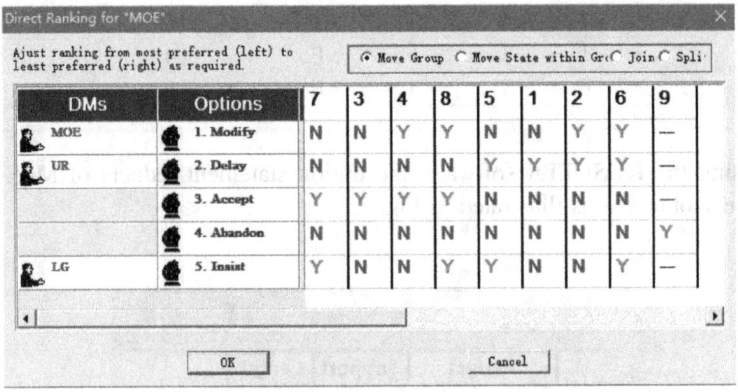

Fig. 4. Ranking of states for MoE under{−4,3, −1, 5IFF−1}

DMs	Options	7	3	4	8	5	1	2	6	9
MOE	1. Modify	N	N	Y	Y	N	N	Y	Y	—
UR	2. Delay	N	N	N	N	Y	Y	Y	Y	—
	3. Accept	Y	Y	Y	Y	N	N	N	N	—
	4. Abandon	N	N	N	N	N	N	N	N	Y
LG	5. Insist	Y	N	N	Y	Y	N	N	Y	—

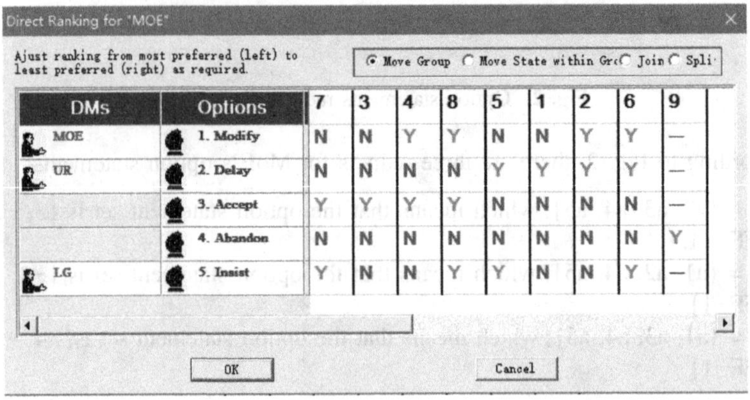

Fig. 5. Ranking of states for MoE under {−4, −2, −1, 5IFF−1}

As displayed in Figs. 3, 4 and 5, the rankings of states for MoE under$\{-4, 3, -1,$ 5IFF$-1\}$ and $\{-4, -2, -1, $5IFF$-1\}$ are the same as MoE's initial order of states with the set of option statements being $\{-4, 3, -2, -1, $5IFF$-1\}$. Alternatively, $\{-4, 3, -1,$ 5IFF$-1\}$ and $\{-4, -2, -1, $5IFF$-1\}$ are the minimal option statement reduct sets of $\{-4, 3, -2, -1, $5IFF$-1\}$ for MoE. Similarly, one can continue to calculate the option statement reducts for UR and LG.

By using the Rosetta software, $\{3$ IFF1$, -5, 2$ IFF$-5\}$ is the reduct for UR's option statements $\{3$ IFF1$, -4, -5, 2$ IFF$- 5\}$. And the ranking of states for UR under the option statement reduct is $s_1 \succ s_4 \succ s_8 \sim s_9 \succ s_5 \succ s_2 \succ s_3 \succ s_7 \succ s_6$, which is not equal to the initial preference of UR. Hence, there are no redundant option statements of UR. For LG's option statements $\{-4, -1, $3IF$-1, $5IF1$, -2, 5\}$, one can find that there are two reducts: $\{-4, -1, -2, 5\}$ and $\{-4, $3IF$-1, -2, 5\}$. But only the first reduct can generate the preference ranking list which is equal to LG's initial ranking of states. Hence, $\{-4, -1, -2, 5\}$is LG's minimal option statement reduct set, in which the option statements "3IF-1" and "5IF1" are redundant and have been removed.

The option statement reduct sets for MoE and LG can be summarized in Table 7.

Table 7. Option statement reduct sets of MoE and LG

DMs	Initial option statement sets	Reduct sets	Redundant statements
MoE	$\{-4, 3, -2, -1, $5IFF$-1\}$	$\{-4, 3, -1, $5IFF$-1\}$	-2
		$\{-4, -2, -1, $5IFF$-1\}$	3
LG	$\{-4, -1, $3IF$-1, $5IF1$, -2, 5\}$	$\{-4, -1, -2, 5\}$	3IF-1, 5IF1

5   Conclusions and Future Work

Within the framework of GMCR, a DM's preference over states can be conveniently acquired in terms of a set of option statements, in which some redundant or unnecessary statements may exist and could decrease the computing efficiency of preference ranking. To eliminate these redundant option statements, a novel option statement reduction method is proposed in this paper based on rough set theory by regarding the option statements and the order of ranked states as being condition and decision attributes, respectively. The case study shows that redundant option statements can be effectively removed by using the option statement reduction approach developed in this research.

In the future, the proposed option statement reduction method can be enhanced by using other effective attribute reduction approaches in rough set theory. Moreover, it can be also extended for handling various kinds of preference such as strength of preference, unknown preference and hybrid preference.

References

1. Kilgour, D.M., Hipel, K.W., Fang, L.: The graph model for conflicts. Automatica **23**, 41–55 (1987)
2. Fang, L., Hipel, K.W., Kilgour, D.M.: Interactive Decision Making: The Graph Model for Conflict Resolution. Wiley, New York (1993)
3. Xu, H., Hipel, K.W., Kilgour, D.M., Fang, L.: Conflict Resolution Using the Graph Model: Strategic Interactions in Competition and Cooperation. Springer, Cham (2018). https://doi.org/10.1007/978-3-319-77670-5
4. Fang, L., Hipel, K.W., Kilgour, D.M., Peng, X.: A decision support system for interactive decision making-part I: model formulation. IEEE Trans. Syst. Man Cybern. Part C (Appl. Rev.) **33**(1), 42–55 (2003)
5. Fang, L., Hipel, K.W., Kilgour, D.M., Peng, X.: A decision support system for interactive decision making-part II: analysis and output interpretation. IEEE Trans. Syst. Man Cybern. Part C (Appl. Rev.) **33**(1), 56–66 (2003)
6. Hamouda, L., Kilgour, D.M., Hipel, K.W.: Strength of preference in graph models for multiple-decision-maker conflicts. Appl. Math. Comput. **179**, 314–327 (2006)
7. Hou, Y., Jiang, Y., Xu, H.: Option prioritization for three-level preference in the graph model for conflict resolution. In: Kamiński, B., Kersten, G.E., Szapiro, T. (eds.) GDN 2015. LNBIP, vol. 218, pp. 269–280. Springer, Cham (2015). https://doi.org/10.1007/978-3-319-19515-5_21
8. Yu, J., Pei, L.L.: Investigation of a brownfield conflict considering the strength of preferences. Int. J. Environ. Res. Public Health **15**(2), 393 (2018)
9. Li, K.W., Hipel, K.W., Kilgour, D.M., Noakes, D.J.: Integrating uncertain preferences into status quo analysis with application to an environmental conflict. Group Decis. Negot. **14**, 461–479 (2005). https://doi.org/10.1007/s10726-005-9003-9
10. Yu, J., Hipel, K.W., Kilgour, D.M., Zhao, M.: Option prioritization for unknown preference. J. Syst. Sci. Syst. Eng. **25**, 39–61 (2016). https://doi.org/10.1007/s11518-015-5282-0
11. Bashar, M.A., Hipel, K.W., Kilgour, D.M., Obeidi, A.: Interval fuzzy preferences in the graph model for conflict resolution. Fuzzy Optim. Decis. Making **17**, 287–315 (2018). https://doi.org/10.1007/s10700-017-9279-7
12. Bashar, M.A., Obeidi, A., Kilgour, D.M., Hipel, K.: Modeling fuzzy and interval fuzzy preferences within a graph model framework. IEEE Trans. Fuzzy Syst. **24**, 765–778 (2016)
13. Bashar, M.A., Kilgour, D.M., Hipel, K.W.: Fuzzy option prioritization for the graph model for conflict resolution. Fuzzy Sets Syst. **26**, 34–48 (2014)
14. Wu, N., Xu, Y., Kilgour, D.M.: Water allocation analysis of the Zhanghe River basin using the graph model for conflict resolution with incomplete fuzzy preferences. Sustainability **11**, 1099 (2019)
15. Kuang, H., Bashar, M.A., Hipel, K.W., Kilgour, D.M.: Grey-based preference in a graph model for conflict resolution with multiple decision makers. IEEE Trans. Syst. Man Cybern.: Syst. **45**, 1254–1267 (2015)
16. Zhao, S., Xu, H.: Grey option prioritization for the graph model for conflict resolution. J. Grey Syst. **29**, 14–25 (2017)
17. Rêgo, L.C., dos Santos, A.M.: Probabilistic preferences in the graph model for conflict resolution. IEEE Trans. Syst. Man Cybern.: Syst. **45**(4), 595–608 (2015)
18. Rêgo, L.C., dos Santos, A.M.: Upper and lower probabilistic preferences in the graph model for conflict resolution. Int. J. Approximate Reasoning **98**, 96–111 (2018)
19. Pawlak, Z.: Rough set theory and its applications to data analysis. Cybern. Syst. **29**(7), 661–688 (1998)

20. Yao, Y., Zhao, Y.: Attribute reduction in decision-theoretic rough set models. Inf. Sci. **178** (17), 3356–3373 (2008)
21. Peng, J.: A decision support system for conflict resolution. Ph.D. thesis, University of Waterloo (1999)
22. Bashar, M.A., Kilgour, D.M., Hipel, K.W.: Fuzzy option prioritization for the graph model for conflict resolution. Fuzzy Sets Syst. **246**, 34–48 (2014)

Alternatives vs. Time – Measuring the Force of Distinct Sources of Bargaining Power

Niklas Dahlen[1]([⊠]) and Tilman Eichstädt[2]

[1] HHL Leipzig Graduate School of Management,
Jahnallee 59, 04109 Leipzig, Germany
niklas.dahlen@hhl.de
[2] BBW Hochschule, Leibnizstraße 11-13, 10625 Berlin, Germany

Abstract. This study aims to deepen the understanding of the drivers of bargaining power in negotiations and in particular the role of best alternatives (BATNA) and time pressure. Previous experimental negotiation research mainly focused on the power of BATNA and the influence of the context on the negotiation outcome, raising the question as to whether BATNA is indeed the only relevant power lever in negotiations. Especially game theorists have shown that time-related costs have a decisive influence on negotiation outcomes. The study proposes a framework to actually measure and compare the relevance and force of different power levers in a simulated distributive buyer-seller negotiation. The results suggest that time pressure can be as influential as an alternative; however, students and professionals seem to react differently to power manipulations. Whereas the student sample was significantly influenced by time pressure but not by alternatives, the opposite could be observed in the professional group. The findings question the common belief that alternatives are the key driver of power in negotiations.

Keywords: Negotiation · Power · Alternatives · Time · Bargaining

1 Introduction

With its interface position in social psychology and economics, negotiation research touches upon many areas of viable research including that of Nobel laureates such as Kahneman and Nash or renowned researchers such as Rubinstein and Ury [24, 38, 70]. While many articles deal with some effects of power, only limited attention is paid to the actual source of this negotiation power. On the whole, most authors agree that negotiation power is the driving force, both with regard to negotiation processes and outcome [27]. Still beside the most prominent source of negotiation power, alternatives, limited attention has been paid to other power sources and their interaction. That said, there is still a lack of systematic research into the drivers of bargaining power, which is reflected by Agndal et al. [2] who state: "A few studies address the issue of power […]." [2, p. 11]. However, recently several authors have looked into the variety of sources of negotiation power and their respective interdependency [6, 18, 27, 37]. At the same time, we are observing a period where negotiation experts are increasingly concerned with the abuse of power negotiations by international politicians.

© Springer Nature Switzerland AG 2020
D. C. Morais et al. (Eds.): GDN 2020, LNBIP 388, pp. 56–72, 2020.
https://doi.org/10.1007/978-3-030-48641-9_5

The Negotiation Journal of the program on negotiation of the Harvard Law School dedicated a whole special issue in early 2019 to the potential impact of President Trumps approach to hard bargaining and the unilateral use of power in negotiations [13]. In this context, Pruitt [64] recognizes the need for a better understanding of the role of time in negotiations.

While power is a multifaceted concept and individual threads of research exist, the basic foundations were laid by French and Raven [26, 68] and Emerson [21]. Within negotiation research, authors have contributed in different fields with very specific concepts such as Rubinstein [59, 70], Nash in game theory [54, 55] and Fisher and Ury in conflict resolution [24]. Howard Raiffa is regarded to be the first to establish a comprehensive and cross-disciplinary approach to negotiation research [65, 66] where he identified time and information as potential sources of power alongside the importance of alternatives. Building upon this, Eichstädt et al. [18] very recently provided a first approach to compare drivers of negotiation power. In parallel, Galinsky et al. [27] added status and social capital to the list of potential negotiation power drivers.

Many researchers state that negotiation power is solely defined by the "best alternative to a negotiated agreement", the so-called BATNA. The simple intuition here is that the better one's alternatives, the better one's position of power in a negotiation [71]. The obvious appeal of this concept has resulted in a certain lack of empirical studies that examine power drivers beyond the BATNA concept, thereby hampering the assessment of the strength of individual power drivers [42, 46]. This is confirmed by Agndal et al. [2] who list only six (1.2%) relevant studies in a meta-analysis of 490 studies on business negotiations. Moreover, the studies do not shed light on the drivers of bargaining power but rather on the effects of having power [2]. Unfortunately, there is not even a reliable framework to measure negotiation power or compare the force of different sources of power. In view of this, the strengths of distinct bargaining power drivers have not been compared to each other systematically [7].

Following the thoughts of Galinsky et al. [27], this paper focuses on comparing two distinct sources of power: alternatives and time. Time plays an especially vital role in most buyer-supplier contracts. Typically, one side has a bigger interest to close the contract and win the business. Maybe it is a buyer to ensure production needs to start soon, or maybe it is a sales agent, trying to ensure he gets his bonus. In Just-in-Time based industries, companies put themselves under the risk of coming under heavy pressure of suppliers, if they threaten to stop supply briefly [63]. In project-based industries, such as the energy sector and wind farm construction, delivery timing forms an essential part of contracts and requires special attention throughout the respective project [15]. The importance of time is also highlighted by professional negotiation advisors who are very successful in negotiation consulting, but do not actively publish in the scientific arena [28, 82]. In order to facilitate the analysis and comparison of power levers in negotiations, we propose a simple concept allowing the magnitude of different power levers to be evaluated by manipulating the levers across otherwise stable experimental settings. Moving towards the empirical study, the paper firstly provides a definition of social power before taking a look at the concepts at hand when it comes to alternatives and time in negotiation research. In doing so, attention is directed at the experimental design of previous studies. Our experimental design aims to incorporate the findings from previous studies while remaining realistic so as not to dilute practical implications. In this way, a dyadic, multi-issue and face-to-face buyer-seller negotiation is simulated. The paper closes with the derivation of theoretical and practical implications and suggests ways to further explore the driving forces of negotiation power.

2 Theoretical Background

Definition of Power. Power is the regulative mechanism of our existence defining the lines of human interaction and thus driving business interactions as well [84]. Max Weber defined power as: "the possibility of imposing one's will upon the behavior of other persons" [86, p. 21]. Keltner et al. [39] sharpen the understanding by saying that it is one's relative ability to change others' attitudes [39, 40]. In particular, the relational aspect of power cannot be stressed enough; it is the cornerstone of Emerson's power dependence theory and also the key to Dahl's concept of power [14, 21]. The first question regarding the origin of power has been discussed extensively in literature and received a comprehensive and continuously evolving framework from French and Raven. However, the effects of various power instruments in negotiations received limited attention, especially in the field of management. Only very recently have Reimann, Shen and Kaufmann [69] looked into the effects of power use in buyer-supplier relationships and the use and effects of applying coercive power [69].

Alternatives and BATNA in Negotiation. Most individuals intuitively agree that having an alternative in a negotiation increases one's power. In his power dependence theory, Emerson laid the groundwork for the BATNA concept as he states that one's power is subject to mutual dependence on each other [21–23]. Consequently, having an alternative unilaterally reduces the dependence of one stakeholder and thus increases his/her bargaining power. A large number of studies have investigated the impact of alternatives in negotiation with different moderating factors pointing to the conclusion that alternatives improve one's position of power. Nevertheless, the variance in experimental setting makes it hard to compare the different and sometimes contradictory results. Moreover, even though the concept of BATNA seems to be rather simplistic, the operationalization is not and Sebenius states that BATNA can be "problematic, limiting and even misleading" [72, p. 1].

Over recent decades, BATNA has been operationalized in many ways but no systematic approach dominates research. These manipulations led to a variety of findings which are ambiguous in some respects [41]. In particular, a lively discussion has formed on the effect of alternatives in terms of the integrativeness of an agreement [89]. Sondak and Bazerman [74] and later Pinkley et al. [62] concluded that an asymmetric power structure leads to a better joint outcome. Both studies involved a job contract negotiation and were conducted with graduate students but their manipulation was slightly different in that three conditions (high, low or no BATNA) were included in the study of Pinkley et al. [62] contrasted with two BATNA conditions in the study of

Sondak and Bazerman [74]. The opposing camp argues that a symmetric power structure leads to a better joint outcome because the frequency of exchange is increased which generates more positive emotions [45, 49].

Multiple studies have dealt with situational factors which can be grouped into three categories: 1) Negotiator's characteristics, 2) Negotiator's decision making, and 3) Other endogenous factors. Studies in the first category involve the impact of emotions [81], social motivation [29, 88], self-efficacy and goal orientation [1, 6]. The studies show that emotions alter concession-making behavior and that an individualistic vs. pro-social motivation leads to more or less competitive negotiation behavior [29, 81]. Having a specific goal or self-efficacy or even both leads to a better individual outcome [1]. Studies on the effect of risky choices [44], choice of negotiation tactic [4] and initiation of the negotiation [47] fall into the second category [43]. Especially note-worthy is the finding of Magee et al. [47] which shows that high-power negotiators are more likely to make the first offer. The final category encompasses endogenous factors that are not directly influenced by the negotiating parties themselves. The studies include the impact of the role of negotiators, size of the bargaining zone, initial offer, and knowledge of the power and quality of BATNA [11, 17, 30, 61, 85]. For the paper at hand, the impact of role is especially interesting and is reviewed by Olekalns [58] who concludes that under equal conditions buyers outperform sellers in terms of profit per transaction, total profit, and number of transactions [5, 20, 57]. Similarly, Neale et al. [36] support this finding by saying that buyers outperform sellers in symmetrical power negotiation in which they manipulated the role information of the parties [36]. Moreover, Eliashberg et al. [20] showed that buyers are perceived as having more power than sellers and reach higher profits in a buyer-seller negotiation on ski caps. In brief, the finding that buyers outperform sellers seems to be robust, as shown by the manipulation of moderating factors such as goal setting or framing [36, 48]. In addition to this, an interesting study by Schaerer et al. [71] demonstrated that having multiple alternatives might actually decrease your individual outcome.

All the above-mentioned studies show that alternatives in negotiation are, at first glance, a well-researched topic. Still, they do not answer the question as to whether and to what extent alternatives are a main driver of bargaining power. Additionally, in the review of Agndal et al. [2] alternatives are neither directly related to power in nego-tiation nor covered in a separate chapter, which indicates that little attention is paid to alternatives as a source of bargaining power.

Time in Negotiation. Even though time dictates the rhythm at which the world operates, it did not receive much attention in negotiation context until 1985. Back then, game theorist Ariel Rubinstein formalized an abstract idea into a concrete concept of either fixed bargaining costs per round or time-dependent costs that reflect the pref-erences of the negotiating parties. In the case of fixed bargaining costs, the parties would arrive at an equilibrium because the party with greater time preference would settle immediately in order to avoid unnecessary costs. In contrast, Rubinstein [70] describes time-dependent costs in terms of a discount factor $\delta_i^t \leq 1$. This would mean that after two negotiation rounds, party 1 would receive $\delta_1^2 x$ and party 2 would receive $\delta_2^2 x$. The party with the lower discount factor therefore has an advantage and can use this against the more impatient party [59, 67, 70].

The cost per round described above is by no means the only trigger of time pressure. Equally well known are approaching deadlines [32, 50, 51, 60]. Additionally, time pressure can arise from threats [73], intervention of third parties [3] or the value of other opportunities for negotiators [12]. While people certainly react differently to time pressure, it is assumed that three strategies to cope with it are applied: 1) Acceleration – To accelerate information processing, 2) Filtration – To only select information perceived as important for processing, and 3) Omission – To use cognitive heuristics and apply a damage minimization strategy [9, 31, 33, 38, 76, 87].

Despite the undisputable importance of time in negotiation research, Rubinstein's pioneering work was neither the starting point for a series of game-theoretical contributions to time pressure in negotiation nor the kick-off of systematic research in another negotiation research field. Looking at the game-theoretical contributions to this topic, Mosterd and Rutte [53] found out that negotiators who act on behalf of somebody else negotiate more competitively under high time pressure. Similarly, Sutter, Kocher and Strauß [78] examined the effect of time pressure on negotiation behavior and saw that the rejection rates increased under high time pressure. They therefore concluded that time pressure in situations that are new to the parties involved leads to losses in efficiency. In the same year, Gneezy, Haruvy and Roth [32] showed that agreements are usually reached at the end of a deadline. While all of the game-theoretical results give us an indication about the general perception of time pressure and its influence on a negotiation, the generalizability of the findings to real-world negotiations is limited due to their specific assumptions (e.g. complete information) and negotiation setting (e.g. ultimatum game).

Non-game-theoretical studies attempted to examine the following: 1) Effect of time pressure and motivational orientation on integrative negotiation [8], 2) Effect of time pressure and information on the negotiation process [77], 3) Effect of time pressure on information processing [16], and 4) Effect of revealing time pressure on the actual outcomes [52]. Even though Carnevale and Lawler [8] and De Dreu [16] applied different negotiation settings, both studies suggest that time pressure and an individualistic orientation lead to more impasses and in general more competitive behavior. Additionally, Stuhlmacher and Champagne [77] demonstrated with their experiment that time pressure reduces the response time and that having additional information about the other party leads to a negotiation advantage [77]. An interesting and at first glance counterintuitive result is provided in studies of Moore [52]. The studies showed that revealing a deadline has a positive impact on the negotiation outcome because the concession making is faster and less time is wasted as a result.

Consequently, the studies on time pressure in negotiation provide valuable insights but they do not offer a definitive answer that can be applied to all scenarios. Additionally, the operationalization of time pressure is not addressed in a concise way, which calls into question the validity of some of the results. The fact that only the study of Moore [52] simultaneously imposed time costs and deadlines is especially troublesome. The lack of systematic research is also documented by the study of Agndal et al. [2] which shows that only nine out of 490 studies involved time pressure in negotiation. The research gap is widened when one considers the limited use of professionals and typical buyer-seller negotiation settings. There is therefore no study that manipulates time pressure in a buyer-seller negotiation with multiple negotiation issues involving students and professionals as participants.

3   Empirical Study

3.1   Method

Negotiation Setting. In order to test the initial hypothesis, a face-to-face negotiation was simulated based on a real-life negotiation on the purchase/sale of corrugated boxes. The experiment includes 50 participants negotiating one on one, of which 68% were male and 32% were female. It was conducted at HHL Leipzig Graduate School of Management and within the scope of an executive training program for HelloFresh AG, which means that a student sample and a professional buyer sample were involved in the experiment. The students were motivated by offering them the possibility to exchange their negotiated outcome for pens of varying quality. The professionals were motivated by both the distribution of chocolate coins based on their performance as well as their fear of losing face in front of their colleagues. The best negotiators were announced publicly and received special recognition of their achievement by the senior management. Additionally, at the beginning of the experiment it was stressed that their performance would be measured based on their individual results. This amplified the individualistic orientation of the participants. In order to reduce the effects of cognitive biases, individuals received an introduction to behavioral economics including the anchoring and framing effect. The parties did not receive any information on their opponents such as age, previous education, etc., as it was assumed that the exposure to the specific negotiation setting was comparable for both negotiation training courses (at HHL Leipzig Graduate School of Management or as part of the executive training program). However, professional buyers in the executive training program had a generally larger exposure to professional negotiations than the students (Professional buyers and students were not mixed in the experiment). Prior to the negotiation, the setup was tested with several dyads with students from HHL Leipzig Graduate School of Management in order to calibrate the time cost and the alternative. Generally, the setting followed an experiment conducted by Eichstädt et al. [18] who applied different power manipulations in executive training programs in the automotive sector. In contrast to the earlier experiments, the following experiment tested more significant manipulations of time costs and compared the negotiation outcome of having increased time costs with having a BATNA.

Negotiation Task. Simulating a real-life negotiation, the participants were randomly assigned to one of two roles (buyer or seller) and had to negotiate the following three issues: price (for 10 pieces in €), payment (days), and minimum order quantity per week. The negotiation was largely based on a real case that was conducted a while ago at HelloFresh. Participants were placed directly opposite one another and were allowed to communicate freely and exchange all the given information. The design aims to be as close as possible to a real negotiation in order to increase the ecological validity and with that the relevance of the implications.

Negotiation Rounds. The experiment extends over three negotiation rounds which lasted a maximum of 15 min and involved time costs amounting to 0.25 points per minute (excluding the third case). The dyads were not mixed in between the rounds. During the different rounds, the following manipulations were applied:

- **1st round:** No manipulation – Parties received information about their own reservation points.
- **2nd round:** Manipulation of alternative – Buyers ould exit the negotiation and take an alternative worth 75% of the available ZOPA. Sellers did not know about the alternative and did not receive one.
- **3rd round:** Manipulation of time pressure – Time costs were doubled for buyers reaching 0.5 points per minute. Time costs for sellers were not changed (0.25 points per minute). Parties received no alternative (BATNA).

Dependent Variable. Performance was solely measured based on the payoff structure, which was determined by the ZOPA and the required time. In other words, the party who is able to claim most of the ZOPA in the shortest amount of time achieves the best outcome. A maximum of 10 points representing 100% of the ZOPA could be reached per round. If the parties did not reach an agreement within 15 min or decided to end the negotiation without an agreement, zero points were awarded. Assuming that a seller claims 60% of the ZOPA within five minutes, an outcome of 4.75 points would be reached (six points from ZOPA – 1.25 points time cost; see Fig. 1). Payoffs were symmetric so that there was no specific integrative solution and the negotiation was purely distributive.

% of ZOPA	Points awarded	Required time	Time costs	Exemplary Outcome
> 10 %	0	1	0.25	
> 20 %	1	2	0.50	
> 30 %	2	3	0.75	
> 40 %	3	4	1.00	
> 50 %	4	5	1.25	>70 % of ZOPA
> 60 %	5	6	1.50	and 5 minutes required
> 70%	6	7	1.75	
> 80%	7	8	2.00	=
> 90 %	8	9	2.25	
> 100%	9	10	2.50	4.75 points
100%	10	11	2.75	
		12	3.00	
		13	3.25	
		14	3.50	
		15	3.75	
			4.00	

Fig. 1. Payoff matrix. Source: Own illustration.

Power Manipulation. In the second round, buyers received an alternative representing 75% of the ZOPA. The fairly good option of receiving 75% of the ZOPA was chosen to ensure that negotiators perceive the alternative as attractive. This manipulation was chosen to give one party substantial leverage stemming from an alternative but still leaving scope to gain a superior result with the opposing party. Additionally, the design ensured easy comprehensibility of the alternative by handing the exact features of the alternative to the participants and not manipulating the likelihood of receiving the alternative [62].

In the third and final round, time costs were altered. Imposing time costs rather than manipulating the deadline was chosen because of three major shortcomings of deadlines. First and foremost, deadlines can only be symmetric which means that if one party stops negotiating then the negotiation ends for both parties [52]. Moreover, deadlines are perceived very differently and thus the resulting behavior varies [50, 75]. In this regard, buyers had to negotiate under twice as much time costs as sellers in this experiment and so 0.5 points per minute were deducted from their result. This approach is based on the reasoning of the Rubinstein model but uses penalties instead of discounting to simplify the decision making for participants. Pretests within the range of manipulation of previous studies were used to calibrate alternatives and time pressure.

3.2 Results

Preliminary Note. The results for the professional and student groups are reported separately using an Analysis of Variance (ANOVA). The effects of the manipulation of time and alternatives are shown in Table 1 and Table 2. These illustrate means of negotiated agreement, role of negotiators, number of dyads, standard deviations, F-statistics, and p-values.

Manipulation Check.[1] After the actual negotiation, participants were asked to fill out a short questionnaire revealing their perceived relative power and time pressure (from "very low" to "very high") across the different cases. The perceived power shows that buyers report significantly higher power in the alternative case than sellers, which indicates a successful manipulation (Buyer: M = 7.63, SD = 1.996; Seller: M = 2.94, SD = 1.731; $F(1, 30) = 50.373$, $p < 0.000$). In the same manner, the results reveal a successful manipulation of time pressure because buyers report significantly higher perceived time pressure in the last case (Buyer: M = 8.88, SD = .885; Seller: M = 2.44, SD = 1.672; $F(1, 30) = 185.256$, $p < 0.000$). Similarly, sellers report higher perceived power in the last case in which buyers are under time pressure (Buyer: M = 3.06, SD = 1.436; Seller: M = 6.25, SD = 1.949; $F(1, 30) = 27.729$, $p < 0.000$).

Effects of Manipulation of Alternatives on Negotiated Outcome. The manipulation of alternatives leads to a better performance of buyers in the student group but the effect is not significant (Seller: M = 2.7656, SD = 4.9273; Buyer: M = 3.2403, SD = 3.2403; $F(1, 30) = 0.550$, $p = 0.464$).

In contrast to this, a significant result can be observed in the professional group. Buyers outperform sellers and score a mean negotiated outcome of 3.69 while sellers only reach 1.44 (Seller: M = 1.4444, SD = 2.1278; Buyer: M = 3.6944,

---

[1] Due to time constraints, no self-assessment was conducted for the executive training program at HelloFresh. Consequently, the manipulation check is limited to the student sample.

SD = 2.6832; F(1, 16) = 3.885, p = 0.066). Additionally, the buyers in both groups performed better having a BATNA compared to the base case.

The findings thus indicate that professionals and students cope differently with the manipulated situation. Professional buyers seem to be able to make better use of alternatives than students.

Effects of Manipulation of Time on Negotiated Outcome. An opposing result in the student and professional groups can be observed here too. Buyers in the student sample who are under severe time pressure are outperformed by their opponent (Seller: M = 4.2813, SD = 2.3235; Buyer: M = 1.9224, SD = 1.9224; F(1, 30) = 5.980, p = 0.021). This does not hold true for the professional sample in which buyers outperform sellers regardless of their time pressure (Seller: M = 2.111, SD = 2.8038; Buyer: M = 3.333, SD = 3.3166; F(1, 16) = 0.713, p = 0.411). Additionally, the sellers in both groups performed better having less time pressure compared to the results of the base case.

Consequently, time pressure has a significant effect on the performance of negotiators in the student sample while professionals seem to be able to counteract the pressure.

Table 1. Results of student sample

Student group (*total of 32 participants or 16 dyads*)				
Manipulation	Mean (standard deviation)		F-ratio	p-value
	Seller	Buyer		
Base	3.9844 (3.52428)	1.5000 (2.03715)	5.960	.21
Alternative favoring the buyer	2.7656 (4.92736)	3.24033 (3.24033)	.550	.464
Time pressure favoring the seller	4.2813 (2.32357)	1.92246 (1.92246)	5.980	.021

Table 2. Results of professional sample

Professional group (*total of 18 participants or 9 dyads*)				
Manipulation	Mean (standard deviation)		F-ratio	p-value
	Seller	Buyer		
Base	1.0833 (1.9243)	2.3611 (2.6900)	1.343	.263
Alternative favoring the buyer	1.4444 (2.1278)	3.6944 (2.6832)	3.885	.066
Time pressure favoring the seller	2.111 (2.8038)	3.333 (3.3166)	.713	.411

4   Discussion

The findings are important, both from a practical and a theoretical perspective. Most significantly they sharpen the understanding of drivers of bargaining power and compare their strength. By doing so, the results provide valuable insights for negotiators into the effects of time pressure and negotiation. Moreover, the experiment shows that students and professionals react differently to power manipulations which highlights the importance of negotiation training. From a theoretical perspective, the paper expands on the thoughts of conceptual papers, for example those of Galinsky et al. [27] and Fleming and Hawes [25]. Additionally, the novel experimental setting contributes to general negotiation research and the ongoing discussion on the validity of student-based experiments.

Practical Implications. The results of previous research have shown that power has a decisive impact on the negotiated result. Additionally, power is a key driver determining many buyer-supplier relationships (BSRs) where we often find a situation where one party (a supplier or buyer) can dominate the relationship based on a better position of power and less dependence [79]. With the identification of two drivers of bargaining power, time pressure and alternatives, the study sensitizes negotiators to the importance of assessing the impact of both on their individual position of power. In fact, an interesting real-life example of managers underestimating the effect of time on the power balance in negotiations could be observed in the German automotive industry in 2016. The relatively small supplier Prevent forced Europe's biggest carmaker Volkswagen to shut down its plants for almost a week by stopping supply after Volkswagen refused to pay for investments in a joint project (VW says 6 plants hit by production stoppages, 2016) [63, 83]. Ultimately, the final agreement was perceived as a bad deal for Volkswagen and a big success for Prevent and shows that negotiation outcomes are not just driven by classical definitions of power like size, economic power and potential to use coercion and reward [69].

Overall, the fact that both time and alternatives influence negotiation power to a certain extent underlines the opportunity for buyers and sellers to develop negotiation strategies, which strengthens their relative power along these lines. The effect of time pressure indicates that managers should pay special attention to the scheduling of negotiations and time management within negotiation. One can create artificial deadlines by defining a specific schedule for the negotiation or scheduling the departure of the negotiation team. While time pressure is perceived by most people as problematic and limiting, it is a two-edged sword which might have an upside. If the negotiation itself creates value for both parties then finding an agreement as fast as possible is beneficial for the two of them [52].

The specific experimental setting involving students and professionals gives us the opportunity to compare their behavior. Students were more affected by the power manipulations which indicates that experience and training are important to withstand an inferior negotiation setting of time pressure or having no alternative. In this way, the study highlights the importance of negotiation training for professional buyers and sellers. The lack of systematic negotiation training is also highlighted in a study by Herbst and Voeth [35] which states that 70% of all respondents did not receive any. In addition, 62% of the surveyed executives stated that they were not prepared for upcoming negotiation tasks by their employer.

Theoretical Implications. In theoretical terms, the most important contribution of the study lies in the investigation of drivers of bargaining power. It is the first study to address the question of which power lever is the strongest. Nonetheless, the findings are not unambiguous since a tendency can be observed that alternatives seem to have an effect in both groups. While students appear to be especially affected by time pressure, professionals can adjust to it and make better use of the provided alternative. This raises the question as to whether experience on the part of the negotiators actually influences the force and impact of negotiation power. The results could indicate that more experienced and professional negotiators benefit from two effects. Firstly, they are able to better use negotiation power with regard to alternatives to their advantage and, secondly, it seems that they are less easily influenced by time pressure to make significant concessions too willingly.

The novelty of the study at hand is also emphasized when looking at the conceptual paper of Galinsky et al. [27]. They state that while there are many studies on the effect of power, there is no common understanding of what power actually is. Based on a literature review, they define alternatives, information, status and social capital as sources of bargaining power. Our study builds on those findings by adding time as a fifth source of bargaining power and gives an empirical foundation for defining alternatives and time as drivers of bargaining power [27]. It also provides a framework to actually measure the relevance of different sources of power, which can additionally be applied to compare those drivers identified by Galinsky et al. [27].

In an article recently published by Fleming and Hawes [25], a negotiation scorecard consisting of 14 situational factors is described which should help to identify an appropriate negotiation strategy. While the authors name timing and power as a determinant of one's strategy, they fail to describe the driving forces behind power and thus complicate the applicability of their concept. Our finding that both time and alternatives are key drivers of bargaining power helps to substantiate the conceptual thoughts and the applicability of the scorecard [25]. Together with a broader assessment of power sources, it could be a useful extension of the scorecard which would help to assess the position of power in advance.

With this considered, the study not only contributes to power in negotiation itself but also to the discussion as to whether student-based experiments are appropriate by applying the same negotiation task to a group of students and professional buyers. Herbst and Schwarz [34] have shown that untrained students are outperformed by experienced managers as well as trained students. While our experiment yields comparable evidence, the novelty lies in the negotiation setting. Instead of an online negotiation which might distort the negotiation results, we applied a face-to-face negotiation. Additionally, we manipulated power levers (time and alternatives) to imitate a real negotiation in which power asymmetries exist. The importance of negotiation experience is also shown by a meta-analysis conducted by ElShenawy [19] which demonstrates that negotiators with extensive negotiation training perform better.

The importance and novelty of this design is also supported by the results of Agndal et al. [2] who point out that only 5.7% of negotiation studies involve both students and professionals. Looking solely at simulations and experiments, only 2.7% (five out of 192) include students and professionals. Moreover, none of the five studies involves a setting in which alternatives and/or time are manipulated.

In addition to this, the findings are important because the experimental design is based on a real negotiation, which gives the participants context. The results thus offer greater relevance compared to experiments based on completely artificial settings [10, 56, 80].

Limitations. Even though the experimental design aimed to incorporate the findings of previous studies, the results must be couched in the caveats of laboratory experiments.

Firstly, the design incorporated only three issues to ensure that the context is easily understandable for everyone. This setting cannot always be observed in reality where buyer-seller negotiations are much more complex and involve a higher number of issues to be negotiated which can hardly be measured on a scale. Secondly, the negotiation was a one-on-one negotiation, meaning that the negotiators were not able to discuss their approach with other individuals. In a typical buyer-seller negotiation, the parties consist of multiple individuals and often multiple rounds, so the decision making might be altered.

Furthermore, the magnitude of calibration of the power manipulation can be discussed because for both manipulations strong empirical evidence or systematic research is missing. As the time manipulation depends on whether a party actually perceives time pressure, it is especially cumbersome to manipulate it in a way that all participants perceive time pressure. Additionally, it could be insightful to include a more thorough analysis of the influence of time pressure on satisfaction or the perceived relationship. Finally, over the course of three rounds of negotiations, learning effects might occur which distort the behavior of the later rounds compared to that of the first.

Future Research. This study is a stepping stone towards a systematic investigation of the different power levers in negotiations. In order to improve the validity of results, several measures need to be taken. Firstly, the experiment should be extended to a higher number of participants from both groups of students and professionals. In addition, it would be recommended to actually measure and quantify the amount of experience, status and knowledge the professionals have in order to test if this alters the impact of negotiation power in line with the findings of Galinsky et al. [27]. Secondly, research on the perception of time pressure and alternatives should be intensified in order to derive a comprehensive approach and calibrate those drivers for research. Thirdly, we have to keep in mind that buyer-seller negotiations always involve humans and the perception of power is thus central to understanding power. Additional research on perception could clarify the underlying cognitive mechanism and help negotiators in their preparations. Finally, additional research should broaden the consideration of other power drivers such as information or social capital to examine their effect and relative strength. The methodology laid out could be easily adapted to measure the impact of sources of negotiation power as put forth by Galinsky et al. [27].

References

1. Adair, W.L., Brett, J.M.: The negotiation dance: time, culture, and behavioral sequences in negotiation. Organ. Sci. **16**(1), 33–51 (2005). https://doi.org/10.1287/orsc.1040.0102
2. Agndal, H., Åge, L., Eklinder-Frick, J.: Two decades of business negotiation research: an overview and suggestions for future studies. J. Bus. Ind. Mark. **32**(4), 487–504 (2007). https://doi.org/10.1108/JBIM-11-2015-0233
3. Arunachalam, V., Dilla, W., Shelley, M., Chan, C.: Market alternatives, third party intervention, and third party informedness in negotiation. Group Decis. Negot. **7**(2), 81–107 (1998). https://doi.org/10.1023/A:1008606709761
4. Bacharach, S.B., Lawler, E.J.: Power and tactics in bargaining power and tactics in bargaining. Labor Relat. Rev. **34**(2), 219–233 (1981)
5. Bazerman, M.H., Magliozzi, T., Neale, M.A.: Integrative bargaining in a competitive market. Organ. Behav. Hum. Decis. Process. **35**(3), 294–313 (1985). https://doi.org/10.1016/0749-5978(85)90026-3
6. Brett, J.F., Pinkley, R.L., Jackofsyk, E.F.: Alternatives to having a BATNA in dyadic negotiation: the influence of goals, self-efficacy, and alternatives on negotiated outcomes. Int. J. Confl. Manag. **7**(2), 121–138 (1996). https://doi.org/10.1108/eb022778
7. Buelens, M., Van De Woestyne, M., Mestdagh, S., Bouckenooghe, D.: Methodological issues in negotiation research: a state-of-the-art-review. Group Decis. Negot. **17**(4), 321–345 (2008). https://doi.org/10.1007/s10726-007-9097-3
8. Carnevale, P.J.D., Lawler, E.J.: Time pressure and the development of integrative agreements in bilateral negotiations. J. Conflict Resolut. **30**(4), 636–659 (1986)
9. Carenevale, P.J., O'Connor, K.M., McCusker, C.: Time pressure in negotiation and mediation. In: Svenson, O., Maule, A.J. (eds.) Time Pressure and Stress in Human Judgment and Decision Making, pp. 117–127. Springer, Boston (1993). https://doi.org/10.1007/978-1-4757-6846-6_8
10. Carnevale, P.J.D., De Dreu, G.K.W.: Methods of Negotition Reserch. Martinus Nijhoff Publishers, Leiden (2006)
11. Conlon, D.E., Pinkley, R.L., Sawyer, J.E.: Getting something out of nothing: Reaping or resisting the power of a phantom BATNA. In: Ayoko, O.B., Ashkanasy, N.M., Jehn, K.A. (eds.) Handbook of Conflict Management Research. Edward Elgar Publishing, Cheltenham (2014)
12. Cross, J.: A theory of the bargaining process. Am. Econ. Rev. **55**(1), 67–94 (1965). http://www.jstor.org/stable/10.2307/1816177
13. Cutcher-Gershenfeld, J., Druckman, D., Manwaring, M., Menkel-Meadow, C., Waters, N.J.: Negotiation and conflict resolution in the age of Trump. Negot. J. **35**(1), 5–8 (2019)
14. Dahl, R.: The concept of power. Behav. Sci. **2**(3), 201 (1957). https://doi.org/10.1002/bs.3830020303
15. D'Amico, F., Mogre, R., Clarke, S., Lindgreen, A., Hingley, M.: How purchasing and supply management practices affect key success factors: the case of the offshore-wind supply chain. J. Bus. Ind. Mark. **32**(2), 218–226 (2017). https://doi.org/10.1108/JBIM-10-2014-0210
16. De Dreu, C.K.W.: Time pressure and closing of the mind in negotiation. Organ. Behav. Hum. Decis. Process. **91**(2), 280–295 (2003)
17. Dwyer, F., Walker, O.: Bargaining in an asymmetrical power structure. J. Mark. **45**(1), 104–115 (1981)

18. Eichstädt, T., Hotait, A., Dahlen, N.: Bargaining power – measuring it's drivers and consequences in negotiations. Group Decis. Negot. **274**, 89–100 (2017). https://doi.org/10.1007/978-3-319-52624-9_7

19. ElShenawy, E.: Does negotiation training improve negotiators' performance? J. Eur. Ind. Train. **34**(3), 192–210 (2010)

20. Eliashberg, J., LaTour, S.A., Rangaswamy, A., Stern, L.W.: Assessing the predictive accuracy of two utility-based theories in a marketing channel negotiation context. J. Mark. Res. **23**(2), 101–110 (1986). https://doi.org/10.2307/3151657

21. Emerson, R.M.: Power-Dependence Relations. Source Am. Sociol. Rev. **27**(1), 31–41 (1986). https://doi.org/10.2307/2089716

22. Emerson, R.M.: Power-dependence relations: two experiments. Sociometry **27**(3), 282–298 (1964)

23. Emerson, R.M.: Social exchange theory. Ann. Rev. Sociol. **2**, 335–362 (1976)

24. Fisher, R., Ury, W.: Getting to Yes. Random House Business Books, New York (1992)

25. Fleming, D.E., Hawes, J.M.: The negotiation scorecard: a planning tool in business and industrial marketing. J. Bus. Ind. Mark. **30**(4) (2017). http://dx.doi.org/10.1108/JBIM-06-2015-0120

26. French, J.R.P., Raven, B.: The bases of social power. In: Leadership as an Influence Process, pp. 151–157 (1959)

27. Galinsky, A.D., Schaerer, M., Magee, J.C.: The four horsemen of power at the bargaining table. J. Bus. Ind. Mark. **32**(4) (2017). https://doi.org/10.1108/jbim-10-2016-0251

28. Gates, S.: The Negotiation Book. Wiley, Chichester (2016)

29. Giebels, E., De Dreu, C., Vliert, E.: Interdependence in negotiation: effects of exit options and social motive on distributive and integrative negotiation. Eur. J. Soc. Psychol. **30**(2), 255–272 (2000)

30. Giebels, E., De Dreu, C., Vliert, E.: The alternative negotiator as the invisible third at the table: the impact of potency information. Int. J. Confl. Manag. **9**(1), 5–21 (1998)

31. Gino, F., Moore, D.: Why negotiators should reveal their deadlines: disclosing weaknesses can make you stronger. Negot. Confl. Manag. Res. **1**(1), 77–96 (2008)

32. Gneezy, U., Haruvy, E., Roth, A.: Bargaining under a deadline: evidence form the reverse ultimatum game. Games Econ. Behav. **45**(2), 347–368 (2003)

33. Goldstein, W.M., Einhorn, H.J.: Expression theory and the preference reversal phenomena. Psychol. Rev. **94**(2), 236–254 (1987)

34. Herbst, U., Schwarz, S.: How valid is negotiation research based on student sample groups? New insights into a long-standing controversy. Negot. J. **27**(2), 147–170 (2011)

35. Herbst, U., Voeth, M.: So verhandeln Deutsche Manager. Harvard Bus. Manag. **1**, 12–15 (2018)

36. Huber, V.L., Northcraft, G.B., Neale, M.A.: The framing of negotiations: contextual versus task frames. Organ. Behav. Hum. Decis. Process. **39**(2), 228–241 (1987). https://doi.org/10.1016/0749-5978(87)90039-2

37. Janusch, H.: The interaction effects of bargaining power: the interplay between veto power, asymmetric interdependence, reputation, and audience costs. Negot. J. **34**(3), 219–241 (2018)

38. Kahneman, D., Tversky, A.: Prospect theory: an analysis of decision under risk. Econometrica **47**(2), 263–291 (1979)

39. Keltner, D., Gruenfeld, D.H., Anderson, C.: Power, approach, and inhibition. Psychol. Rev. **110**(2), 265–284 (2003). https://doi.org/10.1037/0033-295X.110.2.265

40. Keltner, D., Van Kleef, G.A., Chen, S., Kraus, M.W.: A reciprocal influence model of social power: emerging principles and lines of inquiry. Adv. Exp. Soc. Psychol. **40**(7), 151–192 (2008). https://doi.org/10.1016/S0065-2601(07)00003-2

41. Kim, P.H., Fragale, A.R.: Choosing the path to bargaining power: an empirical comparison of BATNAs and contributions in negotiation. J. Appl. Psychol. **90**(2), 373–381 (2005). https://doi.org/10.1037/0021-9010.90.2.373

42. Kim, P.H., Pinkley, R.L., Fragale, A.R.: Power dynamics in negotiation. Acad. Manag. Rev. **30**(4), 799–822 (2005). https://doi.org/10.5465/AMR.2005.18378879

43. Kipnis, D., Schmidt, S.M.: An influence perspective on bargaining within organizations. In: Bazerman, M.H., Lewicki, R.J. (eds.) Negotiating in Organizations, pp. 179–210. Sage, Beverly Hills (1983)

44. Komorita, S., Lapworth, C., Tumonis, T.: The effects of certain vs. risky alternatives in bargaining. J. Exp. Soc. Psychol. **17**(6), 525–544 (1981)

45. Lawler, E.J., Yoon, J.: Power and the emergence of commitment behavior in negotiated exchange. Am. Sociol. Rev. **58**(4), 465–481 (1993)

46. MacGregor, D.: Time pressure and task adaptation: alternative perspectives on laboratory studies. In: Svenson, O., Maule, A.J. (eds.) Time Pressure and Stress in Human Judgment and Decision Making, pp. 73–82. Plenum Press, New York (1993)

47. Magee, J.C., Galinsky, A.D., Gruenfeld, D.H.: Power, propensity to negotiate, and moving first in competitive interactions. Pers. Soc. Psychol. Bull. **33**(2), 200–212 (2007). https://doi.org/10.1177/0146167206294413

48. McAlister, L., Bazerman, M.H., Fader, P.: Power and goal setting in channel negotiations. J. Mark. Res. **23**(3), 228–236 (1986)

49. Mannix, E.A., Neale, M.A.: Power imbalance and the pattern of exchange in dyadic negotiation. Group Decis. Negot. **2**(2), 119–133 (1993). https://doi.org/10.1007/BF01884767

50. Maule, A.J., Hockey, G.R., Bdzola, L.: Effects of time-pressure on decision-making under uncertainty: changes in affective states and information processing strategy. Acta Psychologica **104**(3), 283–301 (2000). https://doi.org/10.1016/S0001-6918(00)00033-0

51. Maule, A.J., Svenson, O.: Time Pressure and Stress in Human Judgment and Decision Making. Plenum Press, New York (1993). https://doi.org/10.1007/978-1-4757-6846-6

52. Moore, D.A.: The unexpected benefits of final deadlines in negotiation. J. Exp. Soc. Psychol. **40**(1), 121–127 (2004). https://doi.org/10.1016/S0022-1031(03)00090-8

53. Mosterd, I., Rutte, G.: Effects of time pressure and accountability to constituents on negotiation. Int. J. Confl. Manag. **11**(3), 227–247 (2000). https://doi.org/10.1108/eb022841

54. Nash, J.: Two-person cooperative games. Econometrica **21**(1), 128–140 (1953)

55. Nash, J.: The bargaining problem. Econometrica **18**(2), 155–162 (1950)

56. Neale, M.A., Simons, T., Thompson, L.: An evaluation of dependent variables in experimental negotiation studies: impasse rates and Pareto efficiency. Organ. Behav. Hum. Decis. Process. **51**(2), 273–295 (1992)

57. Olekalns, M., Smith, P.L.: Dyadic power profiles: power-contingent strategies for value creation in negotiation. Hum. Commun. Res. **39**(1), 3–20 (2013). https://doi.org/10.1111/j.1468-2958.2012.01440.x

58. Olekalns, M.: The balance of power: effects of role and market forces on negotiated outcomes. J. Appl. Soc. Psychol. **21**(12), 1012–1033 (1991)

59. Osborne, M.J., Rubinstein, A.: Bargaining and markets. Games Econ. Behav. **3** (1990). https://doi.org/10.1016/0899-8256(91)90026-B

60. Pinfari, M.: Time to Agree: Time Pressure and 'Deadline Diplomacy' in Peace Negotiations (2010)

61. Pinkley, R.L., Don, V.: Only the phantom knows: impact of certain, conditional, unspecified and zero alternatives to settlement in dyadic negotiation. In: Academy of Management Annual Meeting Proceedings, pp. 82–86 (1997). https://doi.org/10.5465/AMBPP.1997.4980890

62. Pinkley, R.L., Neale, M.A., Bennett, R.J.: The impact of alternative to settlement in dyadic negotiation. Organ. Behav. Hum. Decis. Process. **57**(1), 97–116 (1994)
63. Pretzlaff, H.: Experten warnen vor Kettenreaktion bei VW. Stuttgarter Zeitung, 22.8.2017, Stuttgarter Zeitung Verlagsgesellschaft mbH (2016)
64. Pruitt, D.G.: What have we learned about negotiation from Donald Trump? Negot. J. **35**(1), 87–91 (2019)
65. Raiffa, H.: The Art and Science of Negotiation. First Harvard University Press, Cambridge (1982)
66. Raiffa, H., Richardson, J., Metcalfe, D.: Negotiation Analysis - The Science and Art of Collaborative Decision Making. First Harvard University Press, Cambridge (2007)
67. Rapoport, A., Erev, I., Zwick, R.: An experimental study of buyer-seller negotiation with one-sided incomplete information and time discounting. Manag. Sci. **41**(3), 377–394 (1995)
68. Raven, B.H., Schwarzwald, J., Koslowsky, M.: Conceptualizing and measuring a power Interaction model of interpersonal influence. J. Appl. Soc. Psychol. **28**(4), 307–332 (1998)
69. Reimann, F., Shen, P., Kaufmann, L.: Effectiveness of power use in buyer-supplier negotiations: the moderating role of negotiator agreeableness. Int. J. Phys. Distrib. Logist. Manag. **46**(10), 932–952 (2016). https://doi.org/10.1108/IJPDLM-11-2015-0278
70. Rubinstein, A.: A bargaining model with incomplete information about time preferences. Econometrica **53**(5), 1151–1172 (1985)
71. Schaerer, M., Loschelder, D.D., Swaab, R.I.: Bargaining zone distortion in negotiations: the elusive power of multiple alternatives. Organ. Behav. Hum. Decis. Process. **137**, 156–171 (2016). https://doi.org/10.1016/j.obhdp.2016.09.001
72. Sebenius, J.K.: BATNAs in negotiation: common errors and three kinds of no? Negot. J. **33** (2), 89–99 (2017). https://doi.org/10.1111/nejo.12176
73. Sinaceur, M., Neale, M.A.: Not all threats are created equal: how implicitness and timing affect the effectiveness of threats in negotiations. Group Decis. Negot. **14**(1), 63–85 (2005). https://doi.org/10.1007/s10726-005-3876-5
74. Sondak, H., Bazerman, M.H.: Power balance and the rationality of outcomes in matching markets. Organ. Behav. Hum. Decis. Process. **50**(1), 1–23 (1991)
75. Svenson, O., Benson, L.: On experimental instructions and the inducement of time pressure behavior. In: Svenson, O., Edland, A. (eds.) Time Pressure and Stress in Human Judgment and Decision Making, pp. 157–165. Springer, Boston (1993). https://doi.org/10.1007/978-1-4757-6846-6_11
76. Stuhlmacher, A.F., Gillespie, T.L., Champagne, M.V.: The Impact of time pressure in negotiation: a meta-analysis. Int. J. Confl. Manag. **9**(2), 97–116 (1998). https://doi.org/10.1108/eb022805
77. Stuhlmacher, A.F., Champagne, M.V.: The impact of time pressure and information on negotiation process and decisions. Group Decis. Negot. **9**(6), 471–491 (2000). https://doi.org/10.1023/A:1008736622709
78. Sutter, M., Kocher, M., Strauß, S.: Bargaining under time pressure in an experimental ultimatum game. Econ. Lett. **81**(3), 341–347 (2003)
79. Tangpong, C., Michalisin, M.D., Traub, R.D., Melcher, A.J.: A review of buyer-supplier relationship typologies: progress, problems, and future directions. J. Bus. Ind. Mark. **30**(2), 153–170 (2015)
80. Teich, J., Korhonen, P., Wallenius, H., Wallenius, J.: Conducting dyadic multiple issue negotiation experiments: Methodological recommendations. Group Decis. Negot. **9**(4), 347–354 (2000). https://doi.org/10.1023/A:1008793819872
81. Van Kleef, G.A., DE Dreu, C., Pietroni, D., Manstead, S.: Power and emotion in negotiation: power moderates the interpersonal effects of anger and happiness on concession making. Eur. J. Soc. Psychol. **36**(4), 557–581 (2006)

82. Voss, C.: Never Split the Difference, p. 113ff. Penguin Random House, London (2016)
83. VW says 6 plants hit by production stoppages (2016). http://europe.autonews.com/article/20160821/ANE/160829995/vw-says-6-plants-hit-by-production-stoppages
84. Wallimann, I., Tatsis, N.C., Zito, G.V.: On Max Weber's definition of power. J. Sociol. **13**(3), 231–235 (1970)
85. White, S., Valley, K., Bazerman, M., Neale, A., Peck, S.: Alternative models of price behavior in dyadic negotiations: market prices reservation prices, and negotiator aspirations. Organ. Behav. Hum. Decis. Process. **57**(3), 430–447 (1994)
86. Weber, M.: Wirtschaft und Gesellschaft. Grundriß der verstehenden Soziologie. [Economy and Society. Layout of the interpretive sociology]. Mohr Siebeck, Tübingen (2002)
87. Weenig, M.W.H., Maarleveld, M.: The impact of time constraint on information search strategies in complex choice tasks. J. Econ. Psychol. **23**(6), 689–702 (2002). https://doi.org/10.1016/S0167-4870(02)00134-4
88. Wei, B., Luo, X.: The impact of power differential and social motivation on negotiation behavior and outcome. Public Pers. Manag. **41**(5), 47–58 (2012)
89. Wolfe, R.J., McGinn, K.L.: Perceived relative power and its influence on negotiations. Group Decis. Negot. **14**(1), 3–20 (2005). https://doi.org/10.1007/s10726-005-3873-8

Preference Modeling for Group Decision and Negotiation

Influence Across Agents and Issues in Combinatorial and Collective Decision-Making

Hang Luo$^{(\boxtimes)}$ (iD)

Peking University, Beijing 100871, China
`hang.luo@pku.edu.cn`

Abstract. We consider settings of combinatorial and collective decision-making where a set of agents make choices on a set of issues in sequence based on their preferences over a set of alternatives for each issue. While agents have their initial preferences on issues, they may influence others and be influenced by others, consequently changing their preferences or choices on these issues in the process of decision-making. Though the influence among multiple agents making decisions on one issue and the dependency (influence) among multiple issues decided by one agent have been fully discussed in previous work, the influence from multiple sources across both agents and issues in the context of combinatorial and collective decision-making has been ignored. In this paper, we proposed a preliminary framework to address the influence transcending multiple agents and multiple issues with two rules: *weighted influence* and *one dominant influence*.

Keywords: Combinatorial and collective decision-making · Influence across agents and issues · Weighted influence · One dominant influence

1 Introduction

We consider settings of combinatorial and collective (namely, multi-issue and multi-agent) decision-making, where a set of agents (a general term that can represent a person or an artificial intelligence in nature and that can represent a decision-maker, a voter, or a game player in function, etc.) make choices pertaining to a set of issues in sequence based on their preferences regarding a set of alternatives for each issue. While agents have their initial preferences on a series of issues, they may interact with each other, be fully motivated to influence others, and, accordingly, be influenced by others, consequently changing their preferences and ultimately their choices on these issues in the process of decision-making [32]. The influence on preferences or choices is achieved via the exchange and diffusion of information among agents and across issues. The information that agents exchange can be from other agents' observable (decision-making) behaviors [32] or from others agents' declared or shared preferences (underlying their choices) [25].

© Springer Nature Switzerland AG 2020
D. C. Morais et al. (Eds.): GDN 2020, LNBIP 388, pp. 75–90, 2020.
https://doi.org/10.1007/978-3-030-48641-9_6

Actually, the interaction and influence among agents while making decisions is quite common in reality, and has been studied by scholars from multiple disciplines [25,26], including computer science and artificial intelligence (particularly multi-agent system) [5,17,25,26,32,33,35,37], economics and management science (particularly decision theory and social network) [7–16,19,20,23,24,34], and even politics [28]. For example, in international politics, the decision-making process of the United Nations (UN) Security Council entails various influences among member states, including both positive influences among allies and negative influences among opponents. Each member state in the UN Security Council has full motivation to convince and influence other member states' votes in order to gain desired voting results and maximize its own state interests. Member states in the same alliance usually support each other, therefore positively influencing each other. For instance, the United Kingdom usually casts the same votes as the United States. However, member states from confronting camps usually oppose each other, thereby negatively influencing each other. For instance, the former Soviet Union[1] and present-day Russia usually veto the draft resolutions proposed by the United States.

Moreover, the dependency (influence) among issues for decision-making is also very common in reality, and has first drawn attention from computer scientist [2,3,18,36,38,39]. When an agent is making decisions on a series of issues, it is natural for him/her to refer to his/her own choices regarding the same or similar issues in the past. Namely, an agent's preferences/choices on later issues are usually dependent on (or affected by) his/her own choices on prior issues. There are both positive and negative dependencies among issues. An agent will positively reference (usually, make the same decision as) his/her satisfactory choices in the past but will negatively reference (usually, make the opposite decision to) his/her regrettable choices from the past. For example, in international politics, the United States always used its veto power on the draft resolutions to punish Israel in the UN Security Council, not only for its own state interests but also for its reputation in the international community and particularly in the mind of its allies (a great power should be constant and trustworthy regarding its attitudes toward critical issues and provide a stable expectation for its allies).

2   Related Works

Most previous works either studied the influence among multiple agents while making decisions on a single issue (usually, in the framework of social networks) [5,8–17,19,20,23–26,34,35,37] or studied the dependency among multiple issues decided by a single agent (typically, using the model of CP-nets, namely, conditional preference networks) [2,3,18,36,38,39]. A few studies have combined the influence among multiple agents and the dependency among multiple issues [32,33] in the same model, but they still just discussed them separately and did not study the influence across both agents and issues. Specifically, this means

---

[1] During the first 10 years of the United Nations, the former Soviet Union representative, Andrei Gromyko, even had the nickname "Mr. No".

that there are **horizontal** influences among agents making decisions on a single issue and **vertical** dependencies among issues decided by a single agent, but there is no **diagonal** influence across multiple agents and multiple issues (as shown in Fig. 1) discussed.

It should be noted that traditionally, the dependency among issues would not be deemed the influence, but in fact, the dependency among issues could be understood as a "special" kind of influence. The dependency among issues means that the preferences/choices of an agent on later issues will be affected by his/her own choices on former issues, in a sense, such "affected" just means "influenced". Based on and improving upon the models of [32,33], we build a framework to model the influence across multiple agents and multiple issues, where agents express their preferences as CP-nets, and influences (dependencies) among agents (issues) are expressed as directed links in networks. Before officially representing this framework, we introduce related works on the study of influence in combinatorial and collective decision-making using Example 1, as follows:

Example 1. (UN Security Council Decision-making) This is a typical example of combinatorial and collective decision-making with influences among agents (member states) and dependencies among issues (draft resolutions).

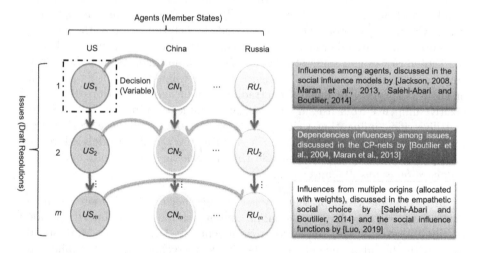

Fig. 1. The UN security council as a typical combinatorial and collective decision-making with influences among agents and dependencies among issues (Notes: Green links represent the dependencies (influences) among issues (draft resolutions) decided by a single agent (member state), and yellow links represent the influences among agents (member states) making decisions on a single issue (draft resolution).) (Color figure online)

First, the UN Security Council decision-making is a typical multi-agent (collective) decision. There are 15 member states (including 5 permanent members and 10 nonpermanent members) collectively making decisions for each draft resolution. Each member state tries to persuade (represented as positively influencing) its allies[2] and opposes (represented as being negatively influenced by) its opponents, in order to pursue its own state interests and achieve desirable voting results on critical international issues. It is easy to see that the UN Security Council is full of varied games, interactions and persuasions among member states, which means that there are influences (expressed as the horizontal directed links shown in Fig. 1) among agents (member states). The models describing the influence among agents are mainly the social influence models [5, 8–17, 19, 20, 25, 26, 35, 37].

Second, the UN Security Council decision-making is also a typical multi-issue (combinatorial) decision. The Council has made decisions on thousands of draft resolutions since the establishment of the UN. Moreover, there are many draft resolutions frequently addressing the same subject (such as the Israel-Palestine issue, Iraq issue and Syria issue). Usually, the votes of a given member state on later draft resolutions would be affected by (refer to) its own votes on former draft resolutions with the same or similar subjects, which means that there are dependencies (represented as the vertical directed links shown in Fig. 1) among issues (draft resolutions). The typical framework describing the dependency among issues is the CP-nets [2, 3, 18, 36, 38, 39].

Third, it should be noted that the influences among member states do not simply flow from one to another but usually simultaneously from multiple ones, which will make the process of influence and the determination of the results of influence complicated. For instance, a vote of China may be influenced by Russia, the United States and some other states at the same time. As such, how do we address the multiple sources of influence and determine the resulting preference or choice, especially in light of contradictive influencing directions (such as a positive influence from Russia and a negative influence from the United States) and diversified influencing weights (such as a stronger influence from a great power and a weaker influence from a small country)? [25] Models addressing multiple sources of influence include the empathetic social choice [37], *social influence functions* [25] and so on.

In fact, influence in present-day society has become much more intensive, particularly with the large-scale online communication via the Internet beyond the limitation of space, time and environment [22–24, 27, 31, 34]. More specifically, with the advancement of wireless network technology and mobile communication devices, particularly with the help of online social platforms, such as Facebook®, Twitter®, and WeChat®, interaction and communication among people (particularly those in remote locations) have become much more convenient and frequent than before [22–24, 27, 31, 34]. If you like, you can instantly communicate with others overseas. Keeping in touch with friends who are far away is no longer a problem in the sense of space, time and environment. Thus, current studies in decision-making should typically involve a very large number of agents interacting with each other and making decisions on a series of issues

---

[2] To let them support what it supports, or oppose what it opposes.

as opposed to restricted cases consisting of a few agents making decisions on a single issue, or a few issues decided by a single agent. In this context, not only should the preference/choice of each single agent for each single issue be discussed, but also the interaction and influence among different agents on different issues should be fully investigated, which makes the study of psychology and behavior of a decision-maker, the mechanisms and dimensions of influence, and the aggregation for collective preference/choice in group decision-making much more complicated.[3]

Moreover, it is not that simple as we might expect even for the influence just among multiple agents (while making decisions on a single issue), as the influence of reality faced by an agent usually comes not from a single agent at a time but from more than one agent at the same time [25, 26]. There are a series of approaches to address multiple sources of influence in group decision-making:

2.1   How to Address Multiple Sources of Influence Among Agents in a Cardinal Approach

Multiple sources of influence in group decision-making have been preliminarily discussed by Salehi-Abari and Boutilier [37] as an empathetic social choice model in the environment of social networks, setting a weight of influence for each influencing agent, in which an agent's utility value is collectively affected by both other agents' utilities and his/her own initial utility. As both the subject of influence and the object of influence are utility values of agents, it is a cardinal (utility value-based) approach. Moreover, it assumes all influence as positive, which may be oversimplifying compared with reality. It is certain that there are positive influences from friends, families, or relatives and so on; however, it is also impossible to avoid all negative influences from enemies, opponents, or any person with a negative appreciation in real-world settings [25].

2.2   How to Address Multiple Sources of Influence Among Agents in an Ordinal Approach

Luo [25] further discussed how to address multiple sources of influence via an ordinal (preference ordering-based) approach and extended the KSB metric [1, 21, 38, 39] to a weighted and signed *matrix influence function* in the context of group decision-making with mutual influence, in which the weight of influence could be stronger or weaker in strength, and positive or negative in polarity.

2.3   From Social Choice Functions to Social Influence Functions

Luo [25] also extended several classical *social choice functions*, including non-ranked choice methods such as the plurality and the majority and ranked

---

[3] Compared with a set of agents independently making choices, combinatorial and collective decision-making with mutual influence entails far more than the simple (linear) aggregation of independent preferences/choices of agents.

choice methods such as the Borda count and the Condorcet method, to signed and weighted *social influence functions*, respectively as: *plurality influence rule, majority influence rule, Borda influence rule* and *Condorcet influence rule*.

However, all these works mainly studied how to address multiple sources of influence among agents' preferences or choices just on a single issue. Overall, the influence from more than one agent making a decision on one issue (namely, the multiple influences in the horizontal dimension) has been fully discussed; besides, the dependency on (influence from) more than one issue for one agent's decision-making (namely, the multiple influences in the vertical dimension) has been preliminarily described in the model of [18]; however, the influence across both more than one agent and more than one issue (namely, the multiple influences in the diagonal dimension) has been ignored.

There is a general meaning for the study of influence across agents and issues in combinatorial and collective decision-making, not only with regard to theory (of computer science, artificial intelligence, social choice and group decision and so on[4]) but also as it pertains to application (of joint-stock company voting, political elections, international organization decision-making[5] and so on).

3   Multiple Sources of Influence Across Agents and Issues

Previous studies have mainly discussed the influence from multiple agents while making decisions on a single issue or the dependency on (influence by) multiple issues for a single agent's decision-making. However, an agent's preference/choice on an issue could be simultaneously influenced by multiple agents' preferences/choices on multiple issues. Thus, the origin of influence is a more general entity, involving both agents and issues. We propose a framework of combinatorial and collective decision-making with influence across multiple agents and multiple issues using Example 2, as follows:

Example 2. (A General Example of Influences across Agents and Issues) Assume a multi-agent and multi-issue decision-making case with a set of agents $\{1, 2, 3\}$ making choices on a set of issues $\{X, Y, Z\}$, each with three alternatives, as shown in Fig. 2. While agent 3 is making a decision on issue Z, it is possible that his/her preference/choice will be influenced by agent 2's preference/choice on the same issue Z (which is an influence between agents making decisions on a single issue) and his/her own preference/choice on former issue Y (which

---

[4] It is specially meaningful in the field of computational social choice [4,6], which is an interdisciplinary research field of computer science (particularly artificial intelligence) and social choice theory. Specifically, it is the study of social choice from the perspective of computer science.

[5] Nearly all international organizations adopt typical group decision-making systems, regardless of whether it is for international political organizations such as the UN General Assembly, UN Security Council [28] and Council of the European Union or international economic organizations such as the World Bank, IMF, Asian Infrastructure Investment Banks [29], and New Development Bank [30].

is a dependency between issues decided by a single agent), while at the same time, influenced by agent 2's preference/choice on former issue Y and agent 1's preference/choice on former issue X (which are influences across both agents and issues).

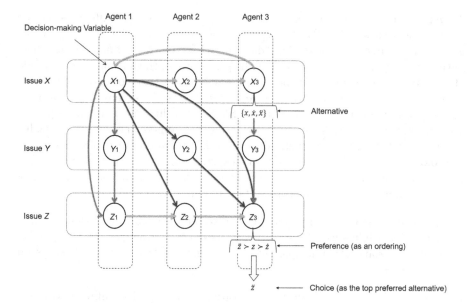

Fig. 2. Influences transcending multiple agents and multiple issues (Notes: Green links represent the dependencies (influences) among multiple issues decided by a single agent; yellow links represent the influences among multiple agents making decisions on a single issue; purple links represent the influences across both agents and issues.) (Color figure online)

When an agent's preference/choice on an issue is simultaneously influenced by more than one agent's preference/choice on more than one issue, especially when each origin of influence has contradictive influencing directions (positive or negative) and varied influencing strengths (weaker or stronger), how to determine the result of the collective influence is an important but not discussed issue. It is relatively easy to set the rule of influence from multiple agents to another agent while making a decision on a single issue or the rule of dependency on multiple (former) issues of another (later) issue decided by a single agent, but it is much more complicated to design a rule of influence from more than one origin across both agents and issues.

In fact, the influence from multiple origins across agents and issues is very common in real-world situations. It is oversimplifying to assume that each agent's preference/choice on each issue is influenced by other agents' preferences/choices only on the same issue, or influenced only by his/her own preferences/choices on other former issues, or only influenced by one other agent's preference/choice on one other issue at a time (or in a round). The reality is that a person's preference/choice on an issue can be influenced by different people (who can be friends, families and so on)'s preferences/choices on different issues to different extents at the same time.

Example 3. (A Specific Example of Stronger and Weaker Influences across Agents and Issues: a Family Buying a Car) Assume a family is choosing a car, and it is a "democratic" instead of "dictatorial" family that all families members have voices. There may be a case that the preference of the husband on the mode of the car will be heavily influenced by his wife's preference or choice regarding the manufacturer of the car: if his wife wants a BMW®, he may be much more inclined to buy a commercial car than a SUV, but if his wife wants a Jeep®, he may be much more inclined to buy a SUV than a commercial car; meanwhile, his preference will be slightly influenced by his kids' preferences or choices regarding the color of the car: if his kids want a black car, he may be a little more inclined to buy a commercial car than a sports car, but if his kids want a red car, he may be a little more inclined to buy a sports car than a commercial car.

Moreover, a person's preference/choice on an issue can also be simultaneously positively influenced by some people (usually as friends)'s and negatively influenced by some other people (usually as enemies)'s preferences/choices on different issues.

Example 4. (A Specific Example of Positive and Negative Influences across Agents and Issues: the UN Security Council Voting) During the process of the UN Security Council voting, there are both positive influences and negative influences among member states on a series of issues (draft resolutions) due to the existence of conflicting state interests and confronting alliances (camps)[6]. The vote of a member state on a draft resolution may be positively influenced by the preferences of its allies on the same draft resolution or the votes of its allies on former relevant draft resolutions (with the same or similar subjects) and negatively influenced by the preferences of its opponents on the same draft resolution or the votes of its opponents on former relevant draft resolutions.

In a multi-agent and multi-issue decision-making context, each influencing and influenced entity, which can be defined as decision-making variables (typically as preference and choice), needs two coordinates (one is the issue-coordinate and the other is the agent-coordinate) to be located, namely, to know which agent is making a decision and on which issue (as shown in Fig. 2). We define the decision-making variables and some background variables in a combinatorial and collective decision-making context as follows:

---

[6] Such as the confrontation between the NATO led by the United States and the Warsaw Pact led by the former Soviet Union in the history and the antagonism between the NATO and Russia and its allies nowadays.

Definition 1. *(Combinatorial and Collective Decision-making Society with Influence Across Agents and Issues) Assume a society* $\mathbb{S} = \{\mathbb{N}, \mathbb{I}, \mathbb{P}, \mathbb{C}, \mathbb{W}\}$*:* $\mathbb{N} = \{1, 2, ..., n\}$ *is the set of all agents;* $\mathbb{I} = \{I_{(1)}, I_{(2)}, ..., I_{(m)}\}$ *is the set of all issues (or features of a multiple-feature issue); then, there are* $n \times m$ *decision-making variables (n agents times with m issues) in total;* $\mathbb{P} = \{P_{(1)}(1), P_{(1)}(2), ..., P_{(1)}(m), \; P_{(2)}(1), P_{(2)}(2), ..., P_{(2)}(m), ..., P_{(n)}(1), P_{(n)}(2), ..., P_{(n)}(m)\}$ *is the set of all agents' preferences (such as preference orderings, utilities, beliefs, opinions, decision-making probabilities, etc.) on all issues;* $\mathbb{C} = \{C_{(1)}(1), C_{(1)}(2), ..., C_{(1)}(m), C_{(2)}(1), \; C_{(2)}(2), ..., C_{(2)}(m), ..., C_{(n)}(1), C_{(n)}(2), ..., C_{(n)}(m)\}$ *is the set of all agents' choices on all issues, namely, counting from agent 1's preference/choice on issue 1 to agent n's preference/choice on issue m, in which* $P_{(i)}(q)$ *represents the preference of agent i on issue q,* $C_{(i)}(q)$ *represents the choice of agent i on issue q* $(i \in \mathbb{N}, q \in \mathbb{I})$*;* $\mathbb{W}$ *is the matrix whose entries are the weights of influence between each of two decision-making variables,* $\mathbb{W} = [w_{(i,j)}(q, h)]$ $(i, j \in \mathbb{N}, q, h \in \mathbb{I})$*, in which* $w_{(i,j)}(q, h)$ *reflects the weight of influence from agent i's preference/choice on issue q to agent j's preference/choice on issue h, the weight value indicates both the strength and polarity of the influence,* $w_{(i,j)}(q, h) > 0$ *means a positive influence,* $w_{(i,j)}(q, h) < 0$ *means a negative influence, and* $w_{(i,j)}(q, h) = 0$ *means there is no influence from agent i's preference/choice on issue q to agent j's preference/choice on issue h; besides, the higher* $|w_{(i,j)}(q, h)|$ *is, the stronger is the influence from agent i's preference/choice on issue q to agent j's preference/choice on issue h.*

In this paper, two preliminary rules: *weighted influence* and *one dominant influence*, addressing the influence from more than one origin across agents and issues, are constructed from different perspectives. We first provide a simple comparison of the two rules when it is for the influence among multiple agents while making decisions just on a single issue, as shown in Example 5, then extend them to a multi-agent and multi-issue decision-making context.

Example 5. (Weighted Influence vs. One Dominant Influence among Agents Making Decisions on a Single Issue) Assume a multi-agent decision-making case with 8 agents making choices on one issue with the set of alternatives as: $\{a, b, c\}$, as shown in Fig. 3. While the agent in the middle is making a decision on this issue, he/she is simultaneously influenced by all other agents possessing various preference orderings with diversified weights of influence, in which some influences are positive, some other influences are negative, and some influences are stronger than others and even with 3 times of strength. However, all influences are from multiple agents but toward a single issue's decision. The comparison of how the two rules address the influence across both agents and issues will be discussed in details in following.

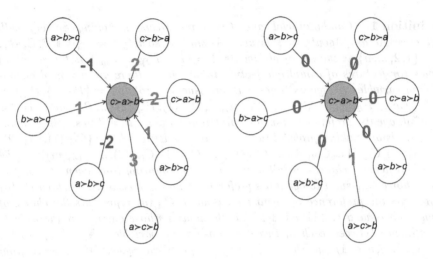

Fig. 3. Weighted influence vs. One dominant influence among multiple agents making decisions on a single issue

4   Weighted Influence Across Agents and Issues

One rule for multiple influences across agents and issues is to assume that the preference/choice (of an agent on an issue) can be collectively influenced by all of the influencing preferences/choices (of more than one agent and on more than one issue), just to different extents and directions according to their respective weights of influence. Actually, the use of the *weight of influence* is a traditional approach to address the influence among agents in group decision-making and has been much discussed in previous works such as [20,25,26,37]. In the empathetic social choice model [37] built via a cardinal approach, the utility value of the influenced agent is the weighted sum of utilities of all influencing agents (including himself/herself and his/her "neighbors"). Luo [25] extended classical *social choice functions*, such as the Borda count and the Condorcet method, to signed and weighted *social influence functions*, which means that both positive and negative influences and both stronger and weaker influences (from all influencing agents) are collectively handled to obtain the resulting choice (of the influenced agent). Moreover, Luo [25] extended the KSB metric [1,21,38,39] to a signed and weighted *matrix influence function* to address the multiple sources of influence among agents via an ordering-based approach; he first defined the rule of how to transform each preference ordering (including both the preference orderings held by the influencing agents and the preference orderings existing in theory)[7] into a matrix and then set a distance metric to compute the distance between any two ordering matrices; the feasible preference ordering that has the smallest weighted sum of distances from all influencing agents' preference

---

[7] If there are m alternatives (candidates), then all the possible preference orderings over them would be $m!$ kinds.

orderings will then be the resulting preference (of the influenced agent) [25, 26]. As the weight of agents' influences can be either positive or negative (namely, as friends or enemies) in real-world settings, it will partially play a role in finding the "closest" possible preference from the positively influencing agents' preferences and partially play a role in finding the "farthest" possible preference from the negatively influencing agents' preferences for the resulting preference [25].

The empathetic social choice [37], *social influence functions* [25] and *matrix influence function* [25] and so on all discussed the influence from more than one agent making a decision on a single issue but not from more than one origin across both agents and issues. However, these models all have potentials to be extended to address the influence transcending multiple agents and multiple issues in the context of combinatorial and collective decision-making. To achieve this, a precondition is to build a weight matrix whose entries are the weights of influence from each decision-making variable $C_{i,q}$ to each of the other decision-making variables $C_{j,h}$ $(i, j \in \mathbb{N}, q, h \in \mathbb{I})$.

Example 6. (A Display of the Weighted Influence across Agents and Issues) Assume a multi-agent and multi-issue decision-making case with a set of agents $\{1, 2, 3\}$ making choices on a set of issues $\{X, Y, Z\}$, each with three alternatives, as shown in Fig. 4. From the perspective of agent 3 making a decision on issue Z, he/she is simultaneously influenced by three agents including himself/herself making decisions on three issues including issue Z. Agent 1 is a friend (ally) in the mind of agent 3, thus, agent 1's preferences/choices on current issue Z and on former issue X produce positive influences (with weights of influence respectively as: 3 and 1) on agent 3's preference/choice on issue Z. Agent 2 is an enemy (opponent) in the mind of agent 3, thus, agent 2's preferences/choices on current issue Z and on former issue Y produce negative influences (with weights of influence respectively as: -2 and -1) on agent 3's preference/choice on issue Z. Furthermore, agent 3's preference/choice on current issue Z is influenced by (dependent on) his/her own preferences/choices on former issue X and Y (with weights of influence respectively as: 3 and 4).

Usually, a choice near current time has higher weight of influence than one other choice far from current time, as people's memories fade with time (regardless of whether it is the satisfaction about a good decision in the past or the regret for a bad decision from the past). What's more, it is quite common that people will be influenced not only by others (such as friends and allies, or enemies and opponents) surrounding them but also by themselves [25]. The latter oneself will inevitably be influenced by the former oneself, and the weight of one's own influence is usually positive [25]. Only in some extreme cases, assume a person encounters serious setbacks and totally loses his/her self-confidence, then his/her own influence could change from strong to weak, and even from positive to negative [25]. Such setup of one's own influence can explain why some people are hard to be influenced by others while some other people are easy to be changed because the former kind of people's self-influences may have higher weights than the latter [25].

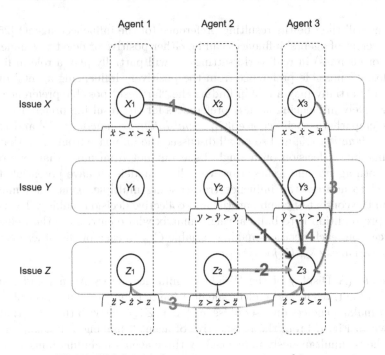

Fig. 4. Weighted influences across agents and issues

5   One Dominant Influence Across Agents and Issues

The other rule for multiple influences across agents and issues is to assume that a preference/choice (of an agent on an issue) will just be influenced by the most predominant preference/choice (with the strongest influencing "power") among all of the influencing preferences/choices (of different agents and on different issues). A concept of the *priority of influence* can be proposed to find the most predominant influencing preference/choice. Assume that there are $n \times m$ decision-making variables (namely, n agents making decisions on m issues) and that there is a priority of influence from each decision-making variable $C_{i,q}$ to each of the other decision-making variables $C_{j,h}$ ($i, j \in \mathbb{N}$, $q, h \in \mathbb{I}$). Thus, each decision-making variable will be influenced by the influencing decision-making variable with the highest priority (of influence) on it compared with all of the other influencing decision-making variables. Actually, one simple method to obtain the priority of influence is to connect it to the weight of influence. Specifically, we make the priority of influence between each of two decision-making variables equal to the absolute value of the weight of influence between them. Thus, for each preference/choice being influenced, the influencing preference/choice that has the highest absolute value of the weight of influence on it also has the highest priority of influence on it and can dominate its result of influence. Intuitively, this rule for addressing multiple sources of influence across agents and issues can remarkably reduce the complexity of the computation.

Moreover, although it is much more simplified compared with *weighted influence*, in a sense, this rule of *one dominant influence* may be closer to how people deal with multiple sources of influence across agents and issues in the real-world. For example, while facing a complicated case of multiple influences from many people making decisions on many issues, you will just follow your best friend (or highest leader) on the most important issue for yourself, or just oppose your most hated enemy on the most critical issue for him/her (only if the absolute value of his/her weight of influence on this issue is the highest, regardless of whether it is positive or negative) rather than engage in a complex weighted deliberation and computation.

Example 7. (A Display of the One Dominant Influence across Agents and Issues) Assume a multi-agent and multi-issue decision-making case with a set of agents $\{1, 2, 3\}$ making choices on a set of issues $\{X, Y, Z\}$, as shown in Fig. 5. Though agent 3 faces complicated multiple influences across both agents and issues while making a decision on issue Z, which is the same case as in Example 6, he/she will just follow the influencing decision-making variable with the highest priority of influence (namely, the highest absolute value of the weight of influence here), which is just the preference/choice of himself/herself on former issue Y. This way to address multiple sources of influence is just like ignoring all other influences except the most predominant one, in mathematics, as if resetting all other influencing preferences/choices' weights to zero (compared with Fig. 4).

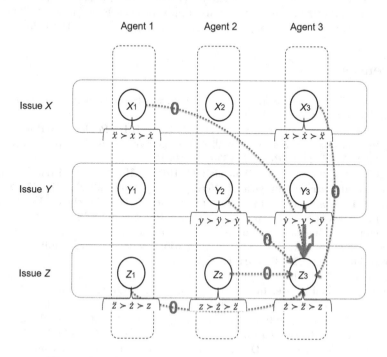

Fig. 5. One dominant influence across agents and issues

6 Discussion and Conclusion

In summary, the study of influence in combinatorial and collective decision-making could be advanced according to the following points:

- We need to discuss group decision-making, not just regarding a singe issue or multiple issues independent to each other but pertaining to multiple issues with combinatorial structures of dependencies among them, constituting a combinatorial and collective decision-making context.
- We should not only discuss the simultaneous influence from multiple agents while making decisions on an issue, but also systematically study the simultaneous dependency on (influence by) multiple issues for an agent's decision-making, particularly investigate the influence from more than one origin across both agents and issues and address the conflicts among multiple sources of influence with varied strengths and contradictive directions.
- To address the multiple influences across agents and issues, two preliminary approaches of *weighted influence* and *one dominant influence* are proposed. In future work, the two approaches could be detailed and improved, and many other promising analytic frameworks and feasible mathematical models could be tried and discussed, particularly considering the case of multiple issues not similar to each other (namely, with different sets of alternatives).

Acknowledgement. This study is supported by a Natural Science Foundation of China Grant (71804006) and a National Natural Science Foundation of China and European Research Council Cooperation and Exchange Grant (7161101045).

References

1. Bogart, K.P.: Preference structures I: distances between transitive preference relations? J. Math. Sociol. **3**(1), 49–67 (1973)
2. Boutilier, C., Brafman, R.I., Domshlak, C., Hoos, H.H., Poole, D.: CP-nets: a tool for representing and reasoning with conditional ceteris paribus preference statements. J. Artif. Intell. Res. **21**, 135–191 (2004)
3. Boutilier, C., Brafman, R.I., Domshlak, C., Hoos, H.H., Poole, D.: Preference-based constrained optimization with CP-Nets. Comput. Intell. **20**(2), 137–157 (2004)
4. Brandt, F., Conitzer, V., Endriss, U.: Computational social choice. In: Weiss, G. (ed.) Multiagent Systems, pp. 213–283. MIT Press, Cambridge (2013)
5. Capuano, N., Chiclana, F., Fujita, H., Herrera-Viedma, E., Loia, V.: Fuzzy group decision making with incomplete information guided by social influence. IEEE Trans. Fuzzy Syst. **26**(3), 1704–1718 (2018)
6. Chevaleyre, Y., Endriss, U., Lang, J., Maudet, N.: A short introduction to computational social choice. In: van Leeuwen, J., Italiano, G.F., van der Hoek, W., Meinel, C., Sack, H., Plášil, F. (eds.) SOFSEM 2007. LNCS, vol. 4362, pp. 51–69. Springer, Heidelberg (2007). https://doi.org/10.1007/978-3-540-69507-3_4
7. Degroot, M.H.: Reaching a consensus. J. Am. Stat. Assoc. **69**(345), 118–121 (1974)
8. Demarzo, P.M., Vayanos, D., Zwiebel, J.: Persuasion bias, social influence, and unidimensional opinions. Q. J. Econ. **118**(3), 909–968 (2003)

9. Friedkin, N.E., Johnsen, E.C.: Social influence and opinions. J. Math. Sociol. **15**(3–4), 193–206 (1990)
10. Friedkin, N.E., Johnsen, E.C.: Social positions in influence networks. Soc. Netw. **19**(3), 209–222 (1997)
11. Golub, B., Jackson, M.O.: Naive learning in social networks and the wisdom of crowds. Am. Econ. J.: Microecon. **2**(1), 112–149 (2010)
12. Grabisch, M., Rusinowska, A.: A model of influence in a social network. Theory Decis. **69**(1), 69–96 (2010). https://doi.org/10.1007/s11238-008-9109-z
13. Grabisch, M., Rusinowska, A.: A model of influence with an ordered set of possible actions. Theory Decis. **69**(4), 635–656 (2010). https://doi.org/10.1007/s11238-009-9150-6
14. Grabisch, M., Rusinowska, A.: Measuring influence in command games. Soc. Choice Welfare **33**(2), 177–209 (2009). https://doi.org/10.1007/s00355-008-0350-8
15. Grabisch, M., Rusinowska, A.: Influence functions, followers and command games. Games Econ. Behav. **72**(1), 123–138 (2011)
16. Grabisch, M., Rusinowska, A.: A model of influence based on aggregation function. Math. Soc. Sci. **66**(3), 316–330 (2013)
17. Grandi, U., Lorini, E., Perrussel, L.: Propositional opinion diffusion. In: Proceedings of the 14th International Conference on Autonomous Agents and Multiagent Systems, pp. 989–997 (2015)
18. Grandi, U., Luo, H., Maudet, N., Rossi, F.: Aggregating CP-nets with unfeasible outcomes. In: O'Sullivan, B. (ed.) CP 2014. LNCS, vol. 8656, pp. 366–381. Springer, Cham (2014). https://doi.org/10.1007/978-3-319-10428-7_28
19. Hoede, C., Bakker, R.: A theory of decisional power. J. Math. Sociol. **8**, 309–322 (1982)
20. Jackson, M.O.: Social and Economic Networks. Princeton University Press, Princeton (2008)
21. Kemeny, J.G., Snell, J.L.: Mathematical Models in the Social Sciences. The MIT Press, Cambridge (1972)
22. Luo, H.: Multi-agent simulation research on urban agglomeration integration and local government interaction. J. Dalian Univ. Technol. **34**(2), 46–52 (2013). in Chinese
23. Luo, H.: Multi-agent system modeling of urban agglomeration's government game and control mechanism: aggregating evolutionary game and small world networks. Chin. J. Syst. Sci. **24**(4), 105–110 (2016). in Chinese
24. Luo, H.: Multi-agent simulation experiment of urban agglomeration's government game and control mechanism: investigate the effect of asymmetrical game situation and complementary compensation measures. Chin. J. Syst. Sci. **25**(2), 78–83 (2017). in Chinese
25. Luo, H.: How to address multiple sources of influence in group decision-making? In: Morais, D.C., Carreras, A., de Almeida, A.T., Vetschera, R. (eds.) GDN 2019. LNBIP, vol. 351, pp. 17–32. Springer, Cham (2019). https://doi.org/10.1007/978-3-030-21711-2_2
26. Luo, H.: Individual, coalitional and structural influence in group decision-making. In: Torra, V., Narukawa, Y., Pasi, G., Viviani, M. (eds.) MDAI 2019. LNCS (LNAI), vol. 11676, pp. 77–91. Springer, Cham (2019). https://doi.org/10.1007/978-3-030-26773-5_7
27. Luo, H., Guo, Z., Zhang, Y.: Value analysis of mobile government. Inf. Doc. Serv. **4**, 36–40 (2010). in Chinese
28. Luo, H., Meng, Q.: Multi-agent simulation of SC reform and national game. World Econ. Polit. **6**, 136–155 (2013). in Chinese

29. Luo, H., Yang, L.: Measuring power in international organizations: from voting weight to voting power: an analysis of AIIB as an example. World Econ. Polit. **2**, 127–154+159–160 (2018). in Chinese

30. Luo, H., Yang, L.: Balance of power and decision-making efficiency in international organizations: an analysis of BRICS new development bank and contingency reserve arrangement as an example. World Econ. Polit. **2**, 123–154+159–160 (2019). in Chinese

31. Luo, H., Zhang, Y., Meng, Q.: Modeling and simulation of multi-cities' policy coordination based on MAS. Chin. J. Manag. Sci. **23**(1), 89–98 (2015). in Chinese

32. Maran, A., Maudet, N., Pini, M.S., Rossi, F., Venable, K.B.: A framework for aggregating influenced CP-nets and its resistance to bribery. In: Proceedings of the Twenty-Seventh AAAI Conference on Artificial Intelligence, pp. 668–674 (2013)

33. Maudet, N., Pini, M.S., Venable, K.B., Rossi, F.: Influence and aggregation of preferences over combinatorial domains. In: Proceedings of the 11th International Conference on Autonomous Agents and Multiagent Systems, pp. 1313–1314 (2012)

34. Meng, Q., Luo, H.: Modeling and simulation of multi-cities' government collaboration based on MAS: embedded in and feedbacking on heterogeneous social networks. J. Manag. Sci. China **20**(3), 183–207 (2017). (in Chinese)

35. Pérez, L.G., Mata, F., Chiclana, F., Gang, K., Herrera-Viedma, E.: Modelling influence in group decision making. Soft Comput. **20**(4), 1653–1665 (2016)

36. Purrington, K., Durfee, E.H.: Making social choices from individuals' CP-nets. In: Proceedings of the 6th International Conference on Autonomous Agent and Multiagent System, p. 179 (2007)

37. Salehi-Abari, A., Boutilier, C.: Empathetic social choice on social networks. In: Proceedings of the 13th International Conference on Autonomous Agents and Multi-Agent Systems, pp. 693–700 (2014)

38. Wicker, A.W., Doyle, J.: Interest-matching comparisons using CP-nets. In: Proceedings of the Twenty-Second AAAI Conference on Artificial Intelligence, pp. 1914–1915 (2007)

39. Wicker, A.W., Doyle, J.: Comparing preferences expressed by CP-networks. In: Proceedings of the AAAI Workshop on Advances in Preference Handling, pp. 128–133 (2008)

A Characterization for Procedural Choice Based on Dichotomous Preferences Over Criteria

Takahiro Suzuki$^{(\boxtimes)}$ and Masahide Horita

Department of International Studies, Graduate School of Frontier Sciences,
The University of Tokyo, 5-1-5, Kashiwanoha, Kashiwashi, Chiba, Japan
ts.takahiro.suzuki@gmail.com, horita@k.u-tokyo.ac.jp

Abstract. Many lessons for procedural choice have been provided by axiomatic studies of decision procedures. However, there appears to be a gap between these axiomatic studies and the actual determination of appropriate procedures, as an axiomatic characterization does not directly answer which axiom should be appropriate—particularly when there is no agreement on the relative desirability of criteria. The present study proposes a formal model of procedural choice based on preferences over criteria (PCBPC). Specifically, we focus on the aggregation rule that maps a dichotomous preference profile over criteria for decision procedures to a nonempty set of decision procedures. We prove that the counting rule, which chooses the decision procedures with greatest supports, is the unique aggregation rule that satisfies anonymity (A), neutrality (N), strict monotonicity (SM), and partition consistency (PC), where PC is proposed based on the idea that representations of equivalent criteria in different ways should not affect the results. Two distinct standpoints for PCBPC are highlighted: one is to regard criteria as atomic, i.e., inseparable, objects and the other as composite, i.e., separable, objects. The difference between them is made explicit with two impossibility theorems showing the inconsistency between unanimity in the former standpoint and A (or PC) in the latter standpoint.

Keywords: Procedural choice · Preferences over criteria · Counting rule

1 Introduction

Axiomatic studies of decision procedures (social welfare functions, social choice rules, multiple criteria decision-making methods (MCDM), etc.) have provided many lessons for aggregating voters' preferences over alternatives. The choice of an appropriate procedure is, however, still a challenging problem and there seems no agreement on the best procedure, even among social choice theorists (Nurmi [1]).

Part of the reason for this difficulty may be that axiomatic characterizations do not, at least in some cases, directly answer which is the appropriate decision procedure. For instance, theorems such as "f is the unique social choice rule that satisfies criteria A and B", or "g is the unique social choice rule that satisfies criteria A and C" deepen our

© Springer Nature Switzerland AG 2020
D. C. Morais et al. (Eds.): GDN 2020, LNBIP 388, pp. 91–103, 2020.
https://doi.org/10.1007/978-3-030-48641-9_7

understanding of these rules. But the choice between f and g is still controversial when there is no agreement on the relative desirability of B and C.

This observation implies the necessity of a decision support system that deals with voters' preferences over criteria. To this end, the present paper provides a formal model for procedural choice based on preferences over criteria (PCBPC). Although each of the axiomatic studies has provided the justifications for particular axioms, Nurmi [1] was the first to study the system for PCBPC (see also de Almeida et al. [2]; application of this method for business contexts is found in de Almeida and Nurmi [3]). Based on the performance matrix of voting rules (Table 1), Nurmi [1] proposes two different methods to create collective rankings/choices of voting rules based on the voters' preferences over criteria. One method is to first determine the rankings of criteria with an ordinary decision procedure (e.g., Borda's or Kemeny's rule) and then determine the rankings of decision procedures (p. 247–248 in Nurmi [1]). The other method is to determine the rankings of decision procedures directly using MCDM methods (p. 250–251 in Nurmi [1]). As Nurmi points out, these methods are designed for situations where voters have linear preferences over criteria and some other methods can be used instead. The present study first aims to provide a formal model for PCBPC in order to make clear the foundations and implicit assumptions of PCBPC. Next, we characterize the counting rule which chooses those decision procedures with the greatest support when voters have dichotomous preferences over criteria.

Table 1. Whether a SCR satisfies a given criterion (1) or not (0) (from de Almeida et al. [2])

	Criterion								
	a	b	c	d	e	f	g	h	i
Amendment	1	1	1	1	0	0	0	0	0
Copeland	1	1	1	1	1	0		0	0
Dodgson	1	0	1	0	1	0	0	0	0
Maximin	1	0	1	1	1	0	0	0	0
Kemeny	1	1	1	1	1	0	0	0	0
Plurality	0	0	1	1	1	1	0	0	1
Borda	0	1	0	1	1	1	0	0	1
Approval	0	0	0	1	0	1	1	0	1
Black	1	1	1	1	1	0	0	0	0
Pl. runoff	0	1	1	0	1	0	0	0	0
Nanson	1	1	1	0	1	0	0	0	0
Hare	0	1	1	0	1	0	0	0	0
Coombs	0	1	1	0	1	0	0	0	0

a: The Condorcet winner criterion, b: The Condorcet loser criterion, c: The strong Condorcet criterion, d: Monotonicity, e: Pareto, f: Consistency, g: Chernoff property, h: Independence of irrelevant alternatives, i: Invulnerability to the no-show paradox. In the table, 1 (0) means that the voting rule satisfies (fails to satisfy) the criterion.

Let us review the literature related to procedural choice. Some authors study the desirable way of procedural choice when the preferences over decision procedures are consequentially induced from the preferences over alternatives (Rae [4]; Koray [5]; Barbera and Jackson [6]; Suzuki and Horita [7]). Others begin with the preferences over decision procedures (Dietrich [8]; Diss and Merlin [9]). Compared with these studies, PCBPC is new in that it is based on the preferences over criteria.

As Nurmi [1] points out, the PCBPC model looks similar to a MCDM model such as PROMETHEE at first glance, because both of them deal with a performance matrix such as Table 1. In our view, the biggest difference between them is that our PCBPC model is based on the social choice theoretic approach: preferences are supposed to be binary relations rather than numerical utility functions; each concept (criterion, aggregation rule, etc.) is defined through set theory; and axioms in social choice theory turn out to be important (as we will argue).

2 Designing a Formal Model for PCBPC

Let $N = \{1, 2, \ldots, n\}$ be the set of voters $(n \geq 2)$ and $\mathcal{F}$ be the set of feasible decision procedures. The elements of $\mathcal{F}$ can be social choice rules, social welfare functions, social decision functions, MCDM methods, etc. A criterion C (for $\mathcal{F}$) is a proposition on the elements of $\mathcal{F}$. More formally, a criterion C is a function that returns "true" or "false" for each $f \in \mathcal{F}$ ($C(f) = $ true means that $f \in \mathcal{F}$ satisfies the criterion).

The most straightforward approach to PCBPC would be simply to use ordinary decision procedures (e.g., Borda's or Kemeny's rule) by regarding criteria as alternatives and then use another procedure to yield the rankings/choices of decision procedures:

Fig. 1. A sketch of a two-step PCBPC

Although this approach is intuitive, it can be controversial because of the differences between criteria and alternatives in the ordinary sense. The most essential difference is in the logical relationships between criteria, i.e., some sets of criteria are consistent, but others are not. As is well known, Arrow [10] 's impossibility theorem says that Pareto, independence of irrelevant alternatives, and non-dictatorship reach a contradiction in the design of social welfare functions (with universal domains). Therefore, to approve of all three criteria equally is logically impossible even though the society members may unanimously agree about them all. On the other hand, the combination of anonymity, neutrality, and positive responsiveness is consistent (May [11]) and so these three can be approved simultaneously. This implies that unsatisfactory results can happen if the two steps are independently determined. In other words, the set of admissible collective judgments of criteria should be restricted by the logical relationships of decision procedures at hand. In order to make matters simple, our model assumes that the PCBPC method returns the collective choices on decision procedures directly from the set of preference profiles over criteria (and not via the collective judgments of criteria).

Assumption 1 (Outline of a PCBPC Method)
A PCBPC method is a single-step procedure[1] that returns a nonempty subset of $\mathcal{F}$ for any profile of voters' preferences over criteria.

When examining voters' preferences over criteria, however, to require such logical consistency seems too demanding. Suppose $\mathcal{F} = \{\text{Amendment}, \text{Copeland}\}$. Then, Table 1 says that criteria a-d are logically equivalent in the sense that f satisfies criterion x if and only if f satisfies criterion y, for all $f \in \mathcal{F}$ and $x, y \in \{a, b, c, d\}$. In this case, logical consistency of voters' preferences over criteria would demand that a: the Condorcet winner criterion and d: Monotonicity must be preferred indifferently. This sounds too demanding especially when the voter is not familiar with those criteria. Assumption 2 allows voters to have any preference irrespective of the consistency:

Assumption 2 (Universal Domain)
The domain, or the set of admissible preference profiles, of the PCBPC method is universal.

Another justification of Assumption 2 is that voters' preferences over criteria can be independent of $\mathcal{F}$. For instance, a voter who favors plurality might prefer the criterion of monotonicity to the Condorcet winner criterion because plurality satisfies only the former. Assumption 2 is necessary for allowing preferences caused by such external factors. In summary, Assumption 2 says that voters may approve even an inconsistent set of criteria (e.g., Pareto, independence of irrelevant alternatives, and non-dictatorship for social welfare functions), and they may approve several criteria independently that are coincidentally equivalent under $\mathcal{F}$.

In the present article, we restrict our attention to dichotomous preferences:

Assumption 3 (Dichotomous Preferences)
Each voter is presumed to have a dichotomous preference over criteria, i.e., each voter either approves or disapproves of each criterion[2].

According to Assumption 3, i's preference over criteria can be expressed by arraying those criteria that i approves, e.g., $(C_{i,1}, C_{i,2}, \ldots, C_{i,a_i})$. For any criterion C, let $\mathcal{F}_C = \{f \in \mathcal{F}/f \text{ satisfies } C\}$.

---

[1] Mathematically speaking, even though the PCBPC method is made up of two separate steps as in Fig. 1, composition of the procedures may turn it into a single-step procedure. In this sense, the phrase "a single-step procedure" in Assumption 1 is not a rigorous mathematical statement, but an intuitive expression of our model.

[2] To make matters simple, we assume that each voter casts their ballot sincerely, i.e., strategic manipulation is assumed not to exist. Thus, we do not distinguish between ballots and preferences.

Assumption 4 (Input to the PCBPC Method: Preferences over Criteria)
The input to the PCBPC method is the preference profile over criteria. A voter's preference over criteria is expressed as a multiset of $\mathfrak{P}(\mathcal{F})$ such as

$$\mathcal{X}_i = \left[\mathcal{F}_{C_{i,1}}, \mathcal{F}_{C_{i,2}}, \ldots, \mathcal{F}_{C_{i,a_i}} \right], \tag{1}$$

where $C_{i,1}, C_{i,2}, \ldots, C_{i,a_i}$ are interpreted as criteria that the voter i approves.
    Let us give some comments on this assumption.

(1) Roughly speaking, Assumption 4 means that the names of the criteria do not matter. Let us explain why. According to Assumption 4, the PCBPC method looks at not the criterion C itself but $\mathcal{F}_C$. In the choices of Copeland and Dodgson, for example, there is supposed to be no difference between criteria b and d in Table 1 because for the two voting rules, satisfying b is equivalent to satisfying d. In this example, Assumption 4 demands that "a voter approves criterion b" is equivalent to "a voter approves criterion d". The point is that Copeland is supported by one reason, no matter which of b or d is approved. In this sense, Assumption 4 is an axiom that treats equivalent criteria equally.

(2) It is also noteworthy that the meaning of a criterion C is measured within the set of feasible decision procedures $\mathcal{F}$. Consequently, the "same" criterion, say Condorcet winner criterion (CW), may be translated into different sets if different $\mathcal{F}$'s are given: for instance, $\mathcal{F}_{CW} = \mathcal{F}$ when $\mathcal{F}$ is made up of amendment, Copeland, and Dodgson, while $\mathcal{F}'_{CW} = \phi$ if $\mathcal{F}'$ is made up of plurality, Borda, and approval (see Table 1). The difference reflects the logical strength of CW in the two cases (CW is extremely weak in the former case and extremely strong in the latter).

In this sense, Assumption 4 demands that the meaning of criteria be argued with respect to $\mathcal{F}$, which can be interpreted as the domain of discourse (what kind of decision procedures are considered). This view is based on the usual practice of social choice theory. One can find many examples where impossibility theorems are conquered by expanding the domain $\mathcal{F}$. One of the most famous cases is that the combination of Pareto, independence of irrelevant alternatives, and non-dictatorship (as well as unrestricted domain) is found to be consistent if social decision functions are considered (Sen [12]). This shows that we cannot argue the consistency without specifying the domain of discourse.

(3) There may be some objections to this assumption. The strongest of them would be "Doesn't it pay little attention to other important features of criteria, such as logical relationships between criteria?" The answer is "yes" and "no". For one thing, the assumption neglects the relationship outside $\mathcal{F}$. The fact that criterion a is stronger than b in the choice of $\mathcal{F}' = \{Amendment, Copeland, Dodgson\}$ is not considered in the choice of $\mathcal{F} = \{Amendment, Copeland\}$ (the two criteria are equivalent under $\mathcal{F}$). At the same time, however, the relationships inside $\mathcal{F}$ are fully included.

This is because the logical relationships between C and C' under $\mathcal{F}$ can be fully expressed by $\mathcal{F}_C$ and $\mathcal{F}_{C'}$[3]

(4) A multiset $[A, B, C, \ldots, Z]$ is a set admitting repetition. Note that the ordering does not matter: $[A, B] \neq [A, B, B] = [B, A, B] = [B, B, A]$.

More formally, we say that $\mathcal{M}$ is a multiset of a set S if $\mathcal{M}$ is a function from S to $\mathbb{Z}_{\geq 0}$ (the set of all nonnegative integers), where $\mathcal{M}(A) = k$ means that A appears k times[4]. For distinction, square brackets are used to express multisets.

(5) We use multisets instead of ordinary sets so that we can admit the approval of some criteria that are coincidentally equivalent under $\mathcal{F}$ (recall Assumption 2).

(6) Note also that our model does not restrict the set of criteria. Each voter can select his/her own set of criteria (e.g., those rules that he/she knows). This is one of the main differences between the present model and Nurmi [1] 's, which considers a fixed set of criteria.

Let $\mathfrak{D}$ be the set of all multisets of $\mathfrak{P}(\mathcal{F})$. Based on Assumption 1-4, we obtain the definition of the PCBPC method (from now on, we call it the aggregation rule).

Definition 1
An *aggregation rule* is a function from $\mathfrak{D}^n$ to the set of all nonempty subsets of $\mathcal{F}$.

3   Results

We use script letters $\mathcal{X}, \mathcal{Y}, \ldots$ to represent preference profiles and script letters with subscripts $\mathcal{X}_i, \mathcal{X}_j, \ldots$ to represent preferences. Subsets of $\mathcal{F}$ are denoted by capital letters $X, Y, \ldots$

Definition 2 (Axioms[5] for Aggregation Rules)
An aggregation rule F is said to satisfy

1) *anonymity* (A) if for all $i, j \in N$ and $\mathcal{X}_i, \mathcal{X}_j \in \mathfrak{D}$ such that $C \in \mathcal{X}_j$, if $\mathcal{X}'_i = \mathcal{X}_i \cup [C]$ and $\mathcal{X}'_j = \mathcal{X}_j \backslash [C]$, then[6] $F(\mathcal{X}_i, \mathcal{X}_j, \mathcal{X}_{-ij}) = F(\mathcal{X}'_i, \mathcal{X}'_j, \mathcal{X}_{-ij})$.

2) *neutrality* (N) if for all $f, g \in \mathcal{F}$, if $\mathcal{X}' \in \mathfrak{D}^n$ is a profile obtained from $\mathcal{X} \in \mathfrak{D}$ by swapping the positions of f and g, then $f \in F(\mathcal{X}) \Leftrightarrow g \in F(\mathcal{X}')$.

---

[3] For instance, "Under $\mathcal{F}$, C is logically stronger than C'" is by definition equivalent to $\mathcal{F}_C \subseteq \mathcal{F}_{C'}$. More complicated statements such as "Under $\mathcal{F}$, if C and not D then E" can be also expressed as $(\mathcal{F}_C \backslash \mathcal{F}_D) \subseteq \mathcal{F}_E$.

[4] Throughout the paper, we use both interpretations of a multiset (i.e., a set with repetition or a function from the base set to $\mathbb{Z}_{\geq 0}$.

[5] Although these are in fact "criteria" for selecting aggregation rules in our sense, we use the word "axiom" in this context to make the distinction from criteria that make up the elements of $\mathcal{F}$.

[6] As usual, preferences other than i and j are expressed as $\mathcal{X}_{-ij}$.

3) *strict monotonicity* (SM) if for all $i \in N$, $\mathcal{X} = (\mathcal{X}_1, \mathcal{X}_2, \ldots, \mathcal{X}_n) \in \mathfrak{D}^n$, and $X \subseteq \mathcal{F}$, if $F(\mathcal{X}) \cap X \neq \phi$, then $F(\mathcal{X}') = F(\mathcal{X}) \cap X$, where $\mathcal{X}'$ is a profile obtained from $\mathcal{X}$ by adding X to someone's preference.
4) *partition consistency* (PC) if for all $i \in N$, disjoint $X, Y \subseteq \mathcal{F}$ such that $X \cup Y \in \mathcal{X}_i$, $F(\mathcal{X}_i, \mathcal{X}_{-i}) = F(\mathcal{X}'_i, \mathcal{X}_{-i})$ where $\mathcal{X}'_i = (\mathcal{X} \setminus [X \cup Y]) \cup [X, Y]$.

As usual, A and N demand that the results are not susceptible to the names of the voters or decision procedures in $\mathcal{F}$, respectively.

SM is also a straightforward modification of strict monotonicity for social choice rules. It demands that if a criterion C is additionally approved by someone, the results will change in favor of the decision procedures supported by the criterion.

The basic idea of PC is that representations of equivalent criteria in different ways should not affect the results. For instance, consider a voter who is in favor of those decision procedures satisfying at least one of either the Condorcet winner criterion (CW) or the Condorcet loser criterion (CL). The union of them can be expressed in several ways[7]:

$$CW \vee CL = (CW \wedge \neg CL) \vee CL = CW \wedge (CL \wedge \neg CW). \tag{2}$$

For this rephrasing, PC demands that each of the following is regarded as the same in the eyes of the aggregation rule:

1) Approve the united axiom $CW \vee CL$;
2) Approve $(CW \wedge \neg CL)$ and CL; and
3) Approve CW and $(CL \wedge \neg CW)$.

Another justification of PC is as follows. Suppose that criterion d in Table 1 (monotonicity) is favored by the society members. At one time, however, the importance of criterion i (invulnerability to the no-show paradox) is argued and the group is divided into three subgroups: [i] those who prefer $d \wedge i$, [ii] those who prefer $d \wedge \neg i$ (maybe those who do not approve criterion i), and [iii] those who prefer both (maybe those who do not care about criterion i). In this case, we can say either that voters in [iii] approve d or that they approve both $(d \wedge i)$ and $(d \wedge \neg i)$. PC demands that these two ways of preference revelation yield the same outcomes.

For a preference profile $\mathcal{X} = (\mathcal{X}_1, \mathcal{X}_2, \ldots, \mathcal{X}_n) \in \mathfrak{D}^n$, we define the appearance of $f \in \mathcal{F}$ at $\mathcal{X}_i$ as

$$a(f : \mathcal{X}_i) := \sum\nolimits_{X \subseteq \mathcal{F} \text{ s.t. } f \in X} \mathcal{X}_i(X). \tag{3}$$

Also, the appearance of $f \in \mathcal{F}$ at $\mathcal{X} = (\mathcal{X}_1, \mathcal{X}_2, \ldots, \mathcal{X}_n)$ is

$$a(f : \mathcal{X}) := \sum\nolimits_{i \in N} a(f : \mathcal{X}_i). \tag{4}$$

---

[7] As usual, $\vee$ means "or" and $\wedge$ means "and".

In words, $a(f : \mathcal{X}_i)$ and $a(f : \mathcal{X})$ represent how many criteria are approved that support f (as noted in (4) of Assumption 4, $\mathcal{X}_i(X)$ denotes how many times X appears in $\mathcal{X}_i$). An aggregation rule F is called the *counting rule*, denoted by F_c, if it selects the decision procedures with the highest appearances:

$$F(\mathcal{X}) = \text{argmax}_{f \in \mathcal{F}} \, a(f : \mathcal{X}). \tag{5}$$

In the later arguments, we use

$$A(\mathcal{X}) = \sum_{f \in \mathcal{F}} a(f : \mathcal{X}). \tag{6}$$

Example 1
Consider a group of three voters, denoted by 1-3. Suppose voter 1 approves criteria a, b, and c in Table 1 (maybe a person who prefers the Condorcet-related criteria) and voter 2 approves e and h (maybe a person who prefers Arrow's framework), and voter 3 approves d and i. Then, among the 13 procedures in Table 1, Copeland, Kemeny, and Black get the highest appearances of 5 pts and thus selected by the counting rule.

Theorem 1
An aggregation rule F satisfies A, N, SM, and PC if and only if it is the counting rule.

[Proof of Theorem 1]
The 'if' part is straightforward. We prove the 'only if' part. Let F be an aggregation rule satisfying A, N, SM, and PC. The proof is made by an induction on $A(\mathcal{X})$.

Suppose that $A(\mathcal{X}) = 0$. For any $f, g \in \mathcal{F}$, $\mathcal{X}$ remains the same when the positions of f and g are swapped. So, N demands that $f \in F(\mathcal{X}) \Leftrightarrow g \in F(\mathcal{X})$ for all $f, g \in \mathcal{F}$. Since $F(\mathcal{X}) \neq \phi$ by definition, it follows that $F(\mathcal{X}) = \mathcal{F}$.

Suppose that the "if" part holds when $A(\mathcal{X}') \leq k$. Suppose $A(\mathcal{X}) = k + 1$. Let $\mathcal{X}^0 = \mathcal{X} = (\mathcal{X}_1, \mathcal{X}_2 \ldots, \mathcal{X}_n)$, $\mathcal{X}_i = [X_{i,1}, X_{i,2}, \ldots, X_{i,a_i}]$, and $E := \{f \in \mathcal{F} | a(f : \mathcal{X}) = k + 1\}$. The proof is completed if we show $F(\mathcal{X}) = E$.

For each $e \in E$, we choose $i(e) \in N$, $j(e) \in \{1, 2, \ldots, a_{i(e)}\}$, and $c(e) = X_{i(e),j(e)}$ such that $e \in X_{i(e),j(e)}$ (such $X_{i(e),j(e)}$ exists because $a(e : \mathcal{X}) = k + 1 > 0$). Let $\mathcal{X}^1$ be a profile obtained from $\mathcal{X}$ by deleting each $e \in E$ from $c(e)$. By the assumption of induction, we have

$$F(\mathcal{X}^1) = \{f \in \mathcal{F} \mid a(f : \mathcal{X}^1) = k\}. \tag{7}$$

Note that $E \subseteq F(\mathcal{X}^1)$. Next, we construct $\mathcal{X}^2$ by adding E to $\mathcal{X}_1^1$ (voter 1's preference at $\mathcal{X}^1$). With SM, we have $F(\mathcal{X}^2) = F(\mathcal{X}^1) \cap E = E$. Let $E = \{e_1, e_2, \ldots, e_k\}$. Divide the added E to singletons in $\mathcal{X}_1^2$ to make $\mathcal{X}_1^3$:

$$\mathcal{X}_1^2 = [\ldots, E] \mapsto \mathcal{X}_1^3 = [\ldots, \{e_1\}, \{e_2\}, \ldots, \{e_k\}]. \tag{8}$$

By repeating PC, we have that $F(\mathcal{X}^3) = F(\mathcal{X}^2) = E$. Construct $\mathcal{X}^4$ by shifting each $\{e\}$ $(e = e_1, e_2, \ldots, e_k)$ from $\mathcal{X}_1^3$ to $\mathcal{X}_{i(e)}^3$. With A, it follows that $F(\mathcal{X}^3) = F(\mathcal{X}^4)$. With PC, we can restore each $X_{i(e),j(e)}$ without changing the results: $F(\mathcal{X}^4) = F(\mathcal{X}^0)$. So, we have $F(\mathcal{X}^0) = E$.   ∎

[Independence of the axioms]

To verify the independence of each axiom used in Theorem 1, we provide several examples. The proof is omitted because they are straightforward.

– A, SM, PC but not N: constant rule $F(\mathcal{X}) = \{f\}$ for some fixed $f \in \mathcal{F}$.
– N, SM, PC, but not A: fix a voter $i^* \in N$. With a slight abuse of notation, let

$$F_C(\mathcal{X}) \cap \mathcal{X}_{i^*} = \{f \in F_C(\mathcal{X}) \mid \exists X \in \mathcal{X}_{i^*} \text{ s.t. } f \in X\}. \tag{9}$$

Define an aggregation rule F as:

$$F(\mathcal{X}) = \begin{cases} F_c(\mathcal{X}) & \text{if } F_C(\mathcal{X}) \cap \mathcal{X}_{i^*} = \phi \\ F_C(\mathcal{X}) \cap \mathcal{X}_{i^*} & \text{otherwise.} \end{cases} \tag{10}$$

– A, N, SM, but not PC: An alternative definition of F_c is as follows. Given a profile $\mathcal{X} = (\mathcal{X}_1, \mathcal{X}_2, \ldots, \mathcal{X}_n)$ with $\mathcal{X}_i = [X_{i,1}, \ldots, X_{i,a_i}]$, F_c searches for each f among $X_{i,1}, X_{i,2}, \ldots, X_{i,a_i}$ and assigns one point to f every time it is found in some $X_{i,j}$. Those decision procedures with the highest points are selected.

Slightly modifying the algorithm above, assign $1/|X_{i,j}|$ points every time f is found in some $X_{i,j}$. Such a rule satisfies A, N, SM, but not PC.

– A, N, PC, but not SM: constant rule $F(\mathcal{X}) = \mathcal{F}$. Another example is the aggregation rule that chooses the decision procedures with the greatest and the second greatest number of appearances.

4 Discussion

In Sect. 3, we provide a characterization of counting rule F_C from the four axioms introduced. In other words, we justify the choice of those decision procedures with highest supports. The significance of the theorem depends on the plausibility of the assumptions as well as the present model. This section discusses what happens if the foundations of Theorem 1 is slightly changed, which clarifies the limitations and the significance of the present model.

(1) *What is a criterion – an atomic element or a composite?*

First, let us discuss the foundation of PC. Let $u(\mathcal{X})$ be the set of criteria that all individuals agree at profile $\mathcal{X}$: $u(\mathcal{X}) = \{X \subseteq \mathcal{F}/X \in \mathcal{X}_i \text{ for all } i \in N\}$.

We say that an aggregation rule satisfies *unanimity on criteria* (UC) if $\mathcal{F}(\mathcal{X}) \subseteq \left(\bigcap_{X \in u(\mathcal{X})} X \right)$ holds whenever $\bigcap_{X \in u(\mathcal{X})} X \neq \phi$. In words, if there exist unanimously agreed criteria and they are consistent in the sense that some decision procedure (s) satisfy them all, the winners must be among such decision procedures. Though it looks plausible, this axiom is inconsistent with PC.

Theorem 2 (Impossibility of PC and UC)
If $|\mathcal{F}| \geq 3$, there is no aggregation rule that satisfies both PC and UC.

(Proof of Theorem 2) Assume to the contrary that F satisfies UC and PC. Let f, g, h be distinct elements in $\mathcal{F}$. Let $\mathcal{X}_1 = [\{f, h\}, \{g, h\}]$ and $\mathcal{X}_i = [\{f\}, \{g\}]$ for all $i \geq 2$. Let $\mathcal{X}_1' = [\{f\}, \{h\}, \{g, h\}]$ and $\mathcal{X}_1'' = [\{f, h\}, \{g\}, \{h\}]$. With PC, we have $F(\mathcal{X}) = F(\mathcal{X}_1', \mathcal{X}_{-1}) = F(\mathcal{X}_1'', \mathcal{X}_{-1})$. But UC implies that $F(\mathcal{X}_1', \mathcal{X}_{-1}) \subseteq \{f\}$ and $F(\mathcal{X}_1'', \mathcal{X}_{-1}) \subseteq \{g\}$. Contradiction. ∎

Readers might find this result somewhat bewildering, because unanimity is one of the most plausible axioms in social choice theory. Thus, one might become concerned with the soundness of PC. Our answer is that these two criteria are based on totally different points of view, which should be called atomic or composite standpoints. Recall that PC demands that different ways to represent a criterion, e.g., $C_1 \vee C_2 = (C_1 \wedge \neg C_2) \vee C_2$, do not affect the results. Consequently, it is possible to partition criteria into pieces without changing the results (see the proof of Theorem 1). Put another way, PC sees each criterion as a composite object that can be decomposed (i.e., a composite standpoint). On the other hand, UC focuses on the form of each criterion as it is, implicitly distinguishing a criterion from its decomposition. In this sense, UC is based on an atomic standpoint: a criterion is indivisible. The impossibility shown in Theorem 2 can be interpreted as the (huge) gap between these two standpoints.

Indeed, the impossibility soon disappears if we think of unanimity based on a composite point of view instead. Let us say that an aggregation rule satisfies *unanimity on procedures* (UP) if whenever there exists a decision procedure f that is unanimously supported by every criterion in the profile, i.e., $f \in X$ for all $X \in \mathcal{X}_1, \mathcal{X}_2, \ldots, \mathcal{X}_n$, then the winners must be found among such f's. One can easily see that UP and PC are consistent, because F_c satisfies them both.

In the rest of this discussion, we argue how PCBPC can be based on the atomic standpoint. To do this, we need to show another impossibility (Theorem 3 below).

(2) *To distinguish criteria that are logically equivalent under $\mathcal{F}$*

Theorem 3 (Impossibility of A and UC)
There is no aggregation rule that satisfies both A and UC.

(Proof of Theorem 3)
    Assume to the contrary that such an aggregation rule F exists. Let $\mathcal{X}_1 = [\{f\}, \{f\}, \{g\}, \{g\}]$, $\mathcal{X}_2 = \phi$, and $\mathcal{X}_i = [\{f\}, \{g\}]$ for all $i \geq 3$. Let $\mathcal{X}_1' = [\{f\}, \{g\}, \{g\}]$,

$\mathcal{X}'_2 = [\{f\}]$, $\mathcal{X}''_1 = [\{f\},\{f\},\{g\}]$, and $\mathcal{X}''_2 = [\{g\}]$. Then, A implies that $F(\mathcal{X}_1, \mathcal{X}_2, \mathcal{X}_{-1,2}) = F(\mathcal{X}'_1, \mathcal{X}'_2, \mathcal{X}_{-1,2}) = F(\mathcal{X}''_1, \mathcal{X}''_2, \mathcal{X}_{-1,2})$. However, UC demands that $F(\mathcal{X}'_1, \mathcal{X}'_2, \mathcal{X}_{-1,2}) \subseteq \{f\}$ and $F(\mathcal{X}''_1, \mathcal{X}''_2, \mathcal{X}_{-1,2}) = \{g\}$. Contradiction. $\blacksquare$

In the proof, two assumptions—the explicit and the implicit—play key roles. The explicit assumption is the very heart of anonymity, i.e., who approves a certain criterion ($\{f\}$ and $\{g\}$ in the proof), must not affect the result. The implicit assumption is the repeated appearances of logically equivalent criteria (two $\{f\}$'s and two $\{g\}$'s are included in a voter's preference), which is in fact the consequence of Assumption 2. As one can see, the proof of Theorem 3 fails if such repetition is not allowed[8]. Therefore, in the last part of this discussion, we argue PCBPC can be based on the atomic standpoint by dropping Assumption 2.

(3) *What if we drop Assumption 2?*

Finally, we give an informal discussion of what happens if we drop Assumption 2. Specifically, we consider a modified model that prohibits us from claiming logically equivalent criteria in a voter's preference. This restriction can be expressed by substituting $\mathbb{Z}_{\geq 0}$ with $\{0, 1\}$ in the co-domain of $\mathcal{X}_i$.

We define Nurmi's aggregation rule F_N below. Note that the rule is slightly modified from Nurmi [1] so that it fits our dichotomous preference model.

Let $\mathcal{X} = (\mathcal{X}_1, \mathcal{X}_2, \ldots, \mathcal{X}_n)$ be a profile and $\mathcal{C}$ be the finite set of criteria at hand. Each voter's preference is supposed to be a subset of $\mathcal{C}$. Let $a(C : \mathcal{X})$ be the number of voters who approve criterion $C \in \mathcal{C}$.

[Step 1] *Determination of collective ranking over $\mathcal{C}$.*
Define an weak ordering[9] $\succsim$ over $\mathcal{C}$ such that for all $C, D \in \mathcal{C}$,

$$C \succsim D \Leftrightarrow a(C : \mathcal{X}) \geq a(D : \mathcal{X}). \tag{11}$$

Suppose that all elements in $\mathcal{C}$ are arrayed from the greatest to the least as $C_1 \sim C_2 \sim \cdots \sim C_k \succsim D_1 \sim D_2 \sim \cdots D_l \succsim E_1 \sim E_2 \sim \cdots \sim E_m \succsim F_1 \sim \cdots$.
[Step 2] *Collective choice over $\mathcal{F}$ based on the ranking generated in Step 1.*

Let $\mathcal{F}'_1 = \mathcal{F}_{C_1} \cap \mathcal{F}_{C_2} \cap \cdots \cap \mathcal{F}_{C_k}$. If $\mathcal{F}'_1$ is empty, $F_N(\mathcal{X}) = \mathcal{X}$. If $\mathcal{F}'_1$ is a singleton, $F(\mathcal{X}) = \mathcal{F}'_1$.

Otherwise, let $\mathcal{F}'_2 = \mathcal{F}'_1 \cap \mathcal{F}_{D_1} \cap \mathcal{F}_{D_2} \cap \cdots \cap \mathcal{F}_{D_l}$. If $\mathcal{F}'_2$ is empty, $F_N(\mathcal{X}) = \mathcal{F}'_1$. If $\mathcal{F}'_2$ is a singleton, $F(\mathcal{X}) = \mathcal{F}'_2$.

Otherwise, let $\mathcal{F}'_3 = \mathcal{F}'_2 \cap \mathcal{F}_{E_1} \cap \mathcal{F}_{E_2} \cap \cdots \cap \mathcal{F}_{E_m}$. If $\mathcal{F}'_3$ is empty, $F_N(\mathcal{X}) = \mathcal{F}'_2$. If $\mathcal{F}'_3$ is a singleton, $F(\mathcal{X}) = \mathcal{F}'_3$. Otherwise, consider $\mathcal{F}'_4$ and the process goes in this way until it finds a singleton (if the intersection of all elements in C includes more than one element, then let $F_N(\mathcal{X})$ be the intersection).

---

[8] Note that the proof of Theorem 2 does not use repetition. Therefore, we may state that the gap between PC and UC is more essential than that between A and UC.

[9] Nurmi considers the domain of linear preferences. As a result, he proposes Borda's or Kemeny's rule for this step.

Note that all voters (decision procedures) are treated equally in the algorithm, which means that F_N satisfies A (N). Furthermore, it also satisfies UC. To see this, note that the basic idea of F_N is to search for a singleton subset of $\mathcal{F}$ by descending from "the greatest" criteria (i.e., the greatest elements of $\mathcal{C}$ with respect to $\gtrsim$) to "the least" criteria. Throughout this process, we have $\mathcal{F}'_1 \supseteq \mathcal{F}'_2 \supseteq \mathcal{F}'_3 \cdots$. As unanimously agreed criteria, if they existed, would belong to $\mathcal{F}'_1$, UC holds under F_N. As we noted earlier, this process is based on a different point of view from the counting rule F_c, or the present paper's framework (Assumptions 1-4). To characterize a multi-step aggregation rule like F_N could be an interesting future topic.

5   Concluding Remarks

The present article provides a formal model for PCBPC, or procedural choice based on preferences over criteria. Theorem 1 characterizes the counting rule, which chooses the decision procedures with the highest number of supports by approved criteria, with four axioms: anonymity, neutrality, strict monotonicity and partition consistency. The first three are well-known and essentially the same as those used in Dietrich [8] 's characterization of "counting rule" for a procedural choice based on preferences over "decision procedures"[10]. As in Dietrich's view, the use of anonymity and neutrality may be justified on the grounds that procedural choice is made only by the judgments within the decision body. Whether certain voters or decision procedures have competence against others should not be decided a priori; that is a matter of collective choice. SM is also important to exclude constant rules or such absurd rules that choose those decision procedures with minimal support.

Our model focuses only on dichotomous preferences (Assumption 3). While this is a typical assumption in voting theory, many studies on MCDM often consider cardinal preferences over criteria. Therefore, to look for anonymous, neutral, strict monotonic, and partition consistent aggregation rules under more complicated preferences over criteria (e.g., linear preferences or even fuzzy preferences) is an interesting topic for future research both from technical and practical points of view.

References

1. Nurmi, H.: The choice of voting rules based on preferences over criteria. In: Kamiński, B., Kersten, Gregory E., Szapiro, T. (eds.) GDN 2015. LNBIP, vol. 218, pp. 241–252. Springer, Cham (2015). https://doi.org/10.1007/978-3-319-19515-5_19
2. de Almeida, A.T., Morais, D.C., Nurmi, H.: Criterion based choice of rules. Systems, Procedures and Voting Rules in Context. AGDN, vol. 9, pp. 57–66. Springer, Cham (2019). https://doi.org/10.1007/978-3-030-30955-8_7

---

[10] Indeed, if the admissible preference is restricted to a unique singleton, our counting rule corresponds with (Dietrich [8])'s.

3. de Almeida, A.T., Nurmi, H.: A framework for aiding the choice of a voting procedure in a business decision context. In: Kamiński, B., Kersten, G.E., Szapiro, T. (eds.) GDN 2015. LNBIP, vol. 218, pp. 211–225. Springer, Cham (2015). https://doi.org/10.1007/978-3-319-19515-5_17

4. Rae, D.W.: Decision-rules and individual values in constitutional choice. Am. Polit. Sci. Rev. **63**(1), 40–56 (1969)

5. Koray, S.: Self-selective social choice functions verify arrow and Gibbard-Satterthwaite theorems. Econometrica **68**(4), 981–996 (2000)

6. Barbera, S., Jackson, M.O.: Choosing how to choose: self-stable majority rules and constitutions. Q. J. Econ. **119**(3), 1011–1048 (2004)

7. Suzuki, T., Horita, M.: Plurality, borda count, or anti-plurality: regress convergence phenomenon in the procedural choice. In: Bajwa, D., Koeszegi, S.T., Vetschera, R. (eds.) GDN 2016. LNBIP, vol. 274, pp. 43–56. Springer, Cham (2017). https://doi.org/10.1007/978-3-319-52624-9_4

8. Dietrich, F.: How to reach legitimate decisions when the procedure is controversial. Soc. Choice Welfare **24**(2), 363–393 (2005)

9. Diss, M., Merlin, V.: On the stability of a triplet of scoring rules. Theor. Decis. **69**(2), 289–316 (2010). https://doi.org/10.1007/s11238-009-9187-6

10. Arrow, K.: Social Choice and Individual Values. Wiley, New York (1951)

11. May, K.O.: A set of independent necessary and sufficient conditions for simple majority decision. Econometrica **20**(4), 680–684 (1952)

12. Sen, A.K.: Collective Choice and Social Welfare. Holden-Day, San Francisco (1970)

Influence Among Preferences and Its Transformation to Behaviors in Groups

An Agent-Based Modeling and Simulation of Fertility Intention and Behavior

Hang Luo, Zhenjie Wang(✉), Shengzi Yang, Hanmo Yang, and Yuke Gong

Peking University, Beijing 100871, China
{hang.luo,zhenjie.wang,ysz,hanmo.yang,
gongyuke}@pku.edu.cn

Abstract. We consider settings of group decision and negotiation where agents' preferences (such as intentions, beliefs and opinions) are influenced by each other and thus their behaviors are possibly changed. We build a multi-agent system (MAS_{IITIB}) in the context of social networks to model the mutual influence among agents' intentions and the transformation from agents' intentions to their behaviors in groups. On the micro level of individual agents, we construct the self-evolution rule of agents' intentions and the generation, constraint and termination rule of agents' behaviors; on the macro level of networked structure, we describe the mutual influence on intentions among agents, which can be diversified in both strength and polarity. We detail this multi-agent system and run experiments and simulations using the interaction of fertility intentions and the generation of fertility behaviors in families as example. Two experimental programs are designed: one is to adjust the initial fertility intentions of prospective parents, and the other is to adjust the range of weight of influence among family members, to investigate the effects on the childbearing behavior and the number of newborn children in the long-term. This study intends to provide modeling bases for the dynamics of preferences and behaviors due to mutual influence among agents in groups, particularly for the study of fertility intention and behavior in families, and more broadly, the forecast of population growth and effects of fertility policy. It is an innovative try of distributed artificial intelligence (multi-agent system) in the field of demography and public policy, and provides with a new bottom-to-up perspective and unconventional agent-based method.

Keywords: Influence among preferences · Transformation from preferences to behaviors · Agent-based modeling and simulation · Fertility intention · Fertility behavior

D. C. Morais et al. (Eds.): GDN 2020, LNBIP 388, pp. 104–119, 2020.
https://doi.org/10.1007/978-3-030-48641-9_8

1 Introduction

In the context of multi-agent systems (such as group decisions and negotiations), the influence among agents' preferences (such as intentions, beliefs, opinions) and behaviors is quite common and has been discussed by scholars from varied disciplines [20, 21], including computer science (particularly artificial intelligence and multi-agent system) [4, 17, 20, 21, 23, 24, 26, 27], economics, decision theory, and social networks [5–9, 12–16]. During the process of a group decision or negotiation, an agent usually has full motivation to influence others' preferences or behaviors with his/her own preferences or behaviors, to make them believe what he/she believes, behave as what he/she behaves, thus increasing the possibility that his/her preference or behavior becomes the mainstream of the group, and gaining more utility and satisfaction [20].

2 Influence on Preferences Among Agents and Its Transformation to Behaviors in Multi-agent Systems

2.1 Transformation from Preferences to Behaviors

In a multi-agent system, though influencing and influenced with each other, every agent's behavior is (eventually) dependent on his/her own preference, no one can be forced to adopt a behavior against his/her own willingness. Actually, this is a fundamental feature of a multi-agent system (such as a group decision and negotiation) where there is no agent as an (absolute) authority who can command and control other agents' behaviors [19]; thus, influencing an agent's preference is the precondition of influencing his/her behavior. Therefore, in the context of a group decision and negotiation, not only the mutual influence among agents' preferences should be discussed, but also the transformation from agents' influenced preferences to their behaviors needs to be investigated. In fact, an influenced (updated) preference will possibly lead to the adoption of a (new) behavior, or not [21]. There is randomness due to bounded rationality and changing environment. The conditions for the generation of behaviors under influence is an important issue that deserves full discussions.

2.2 Weighted and Signed Influence Among Preferences

In real-world settings, the influence faced by one agent usually comes not from one agent at a time, but from more than one agent at the same time [20, 21]. What's more, the multiple influences are usually diversified in both polarity and strength, such as a positive influence from a friend (or ally) versus a negative influence from an enemy (or opponent) [19, 20] and a strong influence from a close friend (or family, or relative) versus a weak influence from an ordinary friend [20, 21]. Even in a group with members close to each other such as a family, influences from different members can also be distinguished by strength and even polarity (as not all relations in a family are harmonious at all times).

Strength of Influence

The strength of influence between two agents can be affected by many factors, for instance: (1) the actual or perceived power (or force, or authority) of the influencer from the perspective of the influenced one, considering that the bigger the power of the influencer is, the greater is the weight of the influence [20]; (2) or how much the influenced one trusts (or believes) the influencer, considering that the higher the trust is, the greater is the weight of the influence [17]; (3) or the similarity of preferences between the influencer and the influenced one, considering that the more similar their preferences are, the greater is the weight of the influence, as people are usually inclined to listen to the opinions that are same and resist the opinions that are different from their own.

Polarity of Influence

The polarity of influence between two agents can also be affected by many factors, for instance: (1) the closeness of relationship between the influencer and the influenced one, considering that the closer the relationship is, the more possible it is for the influencer to exert a positive influence on the influenced one; (2) or the compatibility of objectives (or purpose, or interests) between the influencer and the influenced one, considering that the more contradictive their objectives are, the more possible it is for the influenced one to receive a negative influence from the influencer.

3 An Example of Influence Among Preferences and Behaviors in Groups: Fertility Intention and Behavior in Families

To model the mutual-influence among agents' preferences and the transformation from agents' preferences to their behaviors in groups with details of practical issues, we choose the case of fertility intention and behavior in families. In fact, a family is a very typical group composed of more than one member, with abundant group behaviors and full of constant communications among members. The interactions among family members consist of typical group decisions and negotiations. Particularly in a "democratic" instead of "dictatorial" family, all family members collectively make decisions and ceaselessly negotiate on critical family issues, such as whether and when to bear a child. This kind of decision will profoundly change the life of a family in the long-term, thus, all family members will join the process of decision-making and usually all try their best to convince and influence other members with their own preferences in order to achieve a desirable result. For example, a couple does not want to bear a child, but the couple's parents may have strong intentions to have a grandchild and will try to convince the couple from time to time; on the other side, the couple may resist their parents' "traditional" opinions and even try to sell their own "modern" ideas to their parents. Therefore, the case of fertility intention and behavior in families could be a perfect example to study the mutual influence among preferences and its transformation to behaviors in groups.

4   From Individual Intention, to Individual Behavior and to Social Evolution: A New Approach for Demography

The fertility rate and population growth have always been a critical issue in human society. Particularly in China, families' reproductive behaviors are profoundly affected and constrained by the population and fertility policies. The well-known fertility policy in China was simplified as the "one-child policy", which had been strictly enforced on childbearing behaviors (especially of urban residents) for more than three decades [10, 28]. Under the one-child policy, 36% of the population strictly had only one child; 53% of the population were permitted to have a second child if their first child was a girl; 9.6% of the population were permitted to have two children regardless of their first child's gender; and 1.6% (mainly ethnic minority) had no limit at all [3]. It implies an average of 1.47 children per couple [11], falling far under the international population replacement rate of 2.1 (which means the population of China will be decreasing fast in the future). The tremendous reduction in fertility rate, together with the improved life expectancy and other demographic and socioeconomic factors, has forced China to face a more rapid and serious challenge of population aging than other countries (particularly the so-called "aging before getting rich"[1]).

From January 1st, 2016, China officially abandoned the one-child family planning policy, allowing all couples in China to have two children regardless of their household registration type (i.e., urban or rural hukou), region, ethnicity, and sibship size [25]. The new policy is called the "universal two-child policy". Because of this major change of fertility policy, approximately an additional 90 million women have become eligible to have the second child [31]. The implications of the universal two-child policy are of great public interest and policy concern. However, the fertility intention and actual fertility behavior may significantly differ from what is intended by the policy. A few studies have estimated the long-term effects of this new policy on fertility level and population aging [30, 32]. Till now, the actual fertility rate under the new policy has been much lower than the government's expectation.

Research on population prediction can be traced to Professor King who built a simple mathematical model and calculated manually in 17th century. Then, Malthus put forward the Malthus growth model based on the assumption of a stable population growth rate. Moreover, Verhulst proposed Logistic block population growth model, which could better describe the variation of population growth [2, 18, 29]. Leslie came up with a comprehensive population prediction model, the well-known Leslie Matrix model, containing multiple factors [22]. Subsequently, an innovative population optimal model that accurately describes the dynamic changes of people's ages was developed by Andrew and Meen [1].

However, all the above studies started mainly from a macro perspective of entire system, focusing on a whole country or a whole society, nearly no work has started from a micro perspective of individual interactions, particularly from the perspective of

---

[1] It means that China has faced a serious aging problem which is more common in developed countries, while China still as a developing country is not rich like other aging societies such as Germany and Japan yet.

members in a family, explored how family members' interactions and mutual influences affect couples' childbearing behaviors. This paper introduces an agent-based modeling and simulation approach for the study of fertility behavior and population growth. It is based on a set of intelligent and autonomous agents that can communicate, interact and influence with each other. Each agent represents an individual member of a family and a multi-agent system represents the interaction among family members. In each interaction, every agent's preference (namely, a family member's fertility intention) is influenced by his/her own and other agents' preferences (namely, other family members' fertility intentions). Under certain conditions based on fertility intentions, a fertility behavior of the family will be produced (namely, a child to be born).

5 Mathematical Model: Variables Definition and Rules Design

We build the mathematical model of a multi-agent system with the mutual influence among agents' intentions and the transformation from agents' intentions to their behaviors (MAS_{IITIB}), using the case of the interaction among fertility intentions and the generation of fertility behaviors in families, define variables and design rules in the model as follows:

5.1 Define Variables

$An_{(i)}(t)$ represents agent i (namely, family member i) at time t (namely, t-th mutual influence among family members), $i \in \mathbb{N} = \{0, 1, 2, \ldots, n-1\}$, $\mathbb{N}$ represents the set of all family members (assume there are n members in a family), where 0 means the prospective mother, 1 means the prospective father, and so on; $t = 1, 2, \ldots$ represents the increasing time (the unit of time is set as one month[2]) since the beginning of simulation. The attributes of individual agents, characteristics of relational links (namely, influences) between agents, and the variables of the environment are defined as follows:

(1) $Intention_{(i)}(t)$, abbreviated as $I_{(i)}(t)$, represents family member i's fertility intention at time t. Set $I_{(i)}(t) = x[-1 \leq x \leq 1]^3$, the higher a member's fertility intention is, the more this member inclines to bear a child or hopes his/her family to bear a child. $I_{(i)}(t) < 0$ means he/she does not want his/her family to bear a child, $I_{(i)}(t) > 0$ means he/she expects his/her family to bear a child;

(2) $Initial\text{-}Age_{(0)}$, abbreviated as $age_{(0)}$, represents the prospective mother's initial age at the beginning of simulation. Set $age_{(0)} = x[180 \leq x \leq 588]$, as the childbearing age of women is usually between 15–49 years (namely, 180–588 months) old;

---

[2] The unit of time is set as one month as a woman ovulates one time every month, thus, having one chance to bear a child every month. Besides, one month is long enough for a full communication, interaction and mutual influence among family members.

[3] A random floating number is given between −1 and 1.

(3) *Age-of-Peak-of-Intention*$_{(0)}$, abbreviated as $age^P_{(0)}$. The intention of a woman to bear a child is usually affected by her age. Set $age^P_{(0)} = x[300 \leq x \leq 408]$, as women's fertility intentions usually peak between 25–34 years (namely, 300–408 months) old. When she is younger than certain age, her fertility intention will increase with time, but when she is older than certain age, her fertility intention will decrease with time;

(4) *Behavior*$_{(0)}(t)$, abbreviated as $B_{(0)}(t)$, represents the prospective mother's fertility behavior at time t. Set $B_{(0)}(t) \in \{0, 1\}$, 0 means no child is born at time t, 1 means one child is born at time t;

(5) *weight*$_{(j,i)}(t)$, abbreviated as $w_{(j,i)}(t)$ $(i, j \in \mathbb{N})$, represents the weight of influence from family member j to family member i at time t. Family members' preferences (such as intentions, beliefs and opinions) are usually influenced by each other constantly. The influences can be stronger or weaker in strength and positive or negative in polarity [19–21] due to the complicated relations in families in real-world (we have to admit that not all relations among all family members at all times are harmonious). Set $w_{(j,i)}(t) = x[-1 \leq x \leq 1]$, $w_{(j,i)}(t) < 0$ means a negative (influencing) relationship between two family members (the more the influencer inclines to have a child in his/her family, the less the influenced one inclines), and $w_{(j,i)}(t) > 0$ means a positive (influencing) relationship between two family members (the more the influencer expects to have a child in his/her family, the more the influenced one expects either), and the higher the absolute value of the weight is, the stronger is the influence.

5.2   Design Rules

Intention Self-evolution Function
Assume that the prospective mother's fertility intention will be mainly affected by her age. When she is younger than her age of peak of intention, her fertility intention will increase with the growth of her age; but when she is older than her age of peak of intention, her fertility intention will decrease with the growth of her age.

$$I'_{(0)}(t) = \begin{cases} I_{(0)}(t-1) + x[0 < x \leq y], \ t + age_{(0)} < age^P_{(0)} \\ I_{(0)}(t-1) - x[0 < x \leq y], \ t + age_{(0)} \geq age^P_{(0)} \end{cases} \tag{1}$$

In which, $I'_{(0)}(t)$ means the prospective mother's fertility intention after affected by her ager at time t, $x[0 < x \leq y]$ means a random value between 0–y, y means the maximum of change of fertility intention due to the growth of age[4].

Intention Mutual-influence Function
Assume that family member i's fertility intention for his/her family to bear a child on current time (t) will be influenced by other family members' and his/her own intention

---

[4] We preliminarily set y as 0.1 in the simulation.

on former time $(t-1)$, as it is really hard for people to avoid the influence from their own family members.

$$I_{(i)}(t) = \frac{\sum_{j \in N \setminus \{0\}} w_{(j,i)}(t-1) I_{(j)}(t-1) + w_{(0,i)}(t-1) I'_{(0)}(t)}{\sum_{j \in N \setminus \{0\}} |w_{(j,i)}(t-1)| + |w_{(0,i)}(t-1)|} \quad i = 0, 1, 2, \ldots, n-1$$

(2)

We assume that the prospective mother's fertility intention that is influencing is different from other family members', as it is the prospective mother's fertility intention after affected by her age $(I'_{(0)}(t))$ that is influencing her own and her family' members' fertility intentions.

Behavior Generation Function
Assume when both the prospective mother and prospective father's fertility intentions are high enough, a childbearing behavior will be motivated, and a child will be born. Both prospective mother and father's fertility intentions are normalized values $(x[-1 \leq x \leq 1])$ and are used as probabilities for the generation of a fertility behavior when they are positive. Thus, the higher both of prospective mother and father's fertility intentions are, the more possible it is for them to bear a child.

$$\text{If } I_{(0)}(t) > x[0 \leq x < 1] \wedge I_{(1)}(t) > x[0 \leq x < 1]$$

$$\text{Then } B_{(0)}(t + x[\alpha \leq x \leq \beta]) = B_{(0)}(t) + 1 \tag{3}$$

In which, $x[0 \leq x < 1]$ means a random value between 0–1. Thus, only when both prospective mother and father's fertility intentions are positive (namely, both of them are willing to have a child), there will be a possibility for the formation of a child-bearing behavior. Besides, the fertilization and pregnancy of a child needs certain length of time. The time is assigned by a random value between α–β months[5].

Behavior Constraint Function
Assume when the mother has just given birth to a child, her fertility intention will drop to a negative value, and cannot give birth to the next child within a certain range of time, according to the natural law.

$$\text{If } B_{(0)}(t) > B_{(0)}(t-1)$$

$$\text{Then } I_{(0)}(t) = x[-1 \leq x < 0] \wedge B_{(0)}(t + x[\gamma \leq x \leq \delta]) = B_{(0)}(t) \tag{4}$$

After the birth of a child, the physical recovery and lactation also need certain length of time before the feasibility of the next pregnancy. The time is assigned by a random value between γ–δ months[6].

---

[5] We preliminarily set α as 10 and β as 12 in the simulation.
[6] We preliminarily set γ as 10 and δ as 18 in the simulation.

Behavior Termination Function

Assume when the age of the prospective mother has reached certain months ε,[7] she cannot give birth any more, then the simulation stops.

$$\text{If } t + age_{(0)} > \varepsilon \text{ Then Stop} \tag{5}$$

6 Computer Model

The computer model is constructed based on the above mathematical model (MAS_{IITIB}). The Netlogo[8] is chosen as the implementation platform for the simulation of mutual influence among agents' preferences and the transformation from agents' preferences to their behaviors in groups, particularly the interaction of fertility intentions and the generation of fertility behaviors in families. The 3D evolutional view of the multi-agent system is presented in Fig. 1. We designed a networked graph to describe the relational structure among family members, with every node/agent representing a family member[9] (including prospective parents, prospective parents' parents, prospective parents' relatives and friends[10]) and each link/tie representing an influencing relation between two family members. All agents interact and influence with each other in multi-periods, collectively constructing the dynamics of fertility intentions and behaviors in a family.

About the setting of structure of social networks in a family as shown in Fig. 1, it is not a complete network as a whole because there are n members but there are not necessarily C_n^2 links among them; however, there should be several sub complete networks within the whole network according to common sense, for example: the set of {prospective father, prospective mother, prospective father's father, prospective father's mother, prospective mother's father, prospective mother's mother} should constitute a complete network which is full of all possible links, but it is not necessary to exist a link (namely, an influencing relation) between the prospective mother's relatives, friends and the prospective father.

7 Simulation Experiments and Results

The aim of our experiments is to simulate the mutual influence among fertility intentions and the transformation from fertility intentions to fertility behaviors in a micro individual perspective of family members, different from the conventional macro entire perspective of social systems in Demography, to investigate the long-term effects of the

---

[7] We preliminarily set ε as 588 (namely, 49 years) old.

[8] There are several platforms such as the Swarm, Repast, Ascape, AnyLogic, Matlab for agent-based system modeling and simulation, in which the Netlogo is notable for its ease of use and friendly interface.

[9] We use agents' colors to distinguish family members' genders: a blue agent indicates a male and a pink agent indicates a female.

[10] We assume the influence from a friend is usually positive.

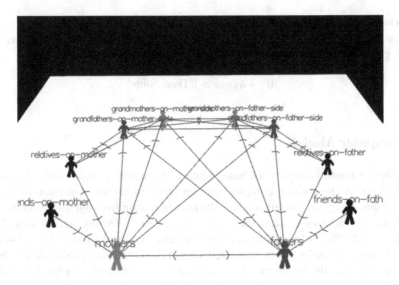

Fig. 1. Networked structure of family members in a household

initial fertility intentions of couples and the relational structure of family members on the fertility rate and population growth, particularly under consideration of the fertility policy changed from the one-child policy to the universal two-child policy in China. We design two experimental programs for computational simulations as follows:

- In experimental program I, we preliminarily set three levels for couples' initial fertility intentions, including: (1) for a low fertility intention, a random value is given from $x[-1 \leq x < 0]$; (2) for a medium fertility intention, a random value is given from $x[-0.5 \leq x \leq 0.5]$; (3) and for a high fertility intention, a random value is given from $x[0 < x \leq 1]$. If both the prospective father and mother's initial fertility intentions have three levels of variations, then the combinations will be $9 = 3 \times 3$, corresponding to 9 experimental plans. We also set a control group where we run the simulation without any intervention on the range of initial fertility intentions, which means a random value is given from $x[-1 \leq x \leq 1]$. Then, we run different experimental plans to simulate their effects on the generations of fertility behaviors and the number of new born children.

- In experimental program II, we set the range of the weight of influence for three categories of relationships between family members, respectively: between the prospective mother and the prospective father of a couple, between a prospective father/mother and his/her own parents, and between a prospective father/mother and his/her parents-in-law. We preliminarily set two levels for each category of relationships, including: (1) for a great relation, a random value of $x[0 \leq x \leq 1]$ is given to the weight of influence, which means the influencing relation must be positive; (2) for a normal relation, a random value of $x[-1 \leq x \leq 1]$ is given to the weight of

Table 1a. Experiment I's outcomes of families' fertility behaviors under adjusting couples' initial fertility intentions

Variable	Experimental outcome: numbers of families (percent of families)				
	Control group	a	b	c	d
No. of children (2 years)					
0	4062 (40.6)	4116 (41.2)	4058 (40.6)	4125 (41.3)	4089 (40.9)
1	3730 (37.3)	3743 (37.4)	3748 (37.5)	3707 (37.1)	3704 (37.0)
2	2208 (22.1)	2141 (21.4)	2194 (21.9)	2168 (21.7)	2207 (22.1)
No. of children (5 years)					
0	3388 (33.9)	3379 (33.8)	3386 (33.9)	3321 (33.2)	3361 (33.6)
1	3058 (30.6)	3181 (31.8)	3135 (31.4)	3228 (32.3)	3184 (31.8)
2	3554 (35.5)	3440 (34.4)	3479 (34.8)	3451 (34.5)	3455 (34.6)
No. of children (lifetime)					
0	193 (1.9)	203 (2.0)	193 (1.9)	199 (2.0)	198 (2.0)
1	1791 (17.9)	1806 (18.1)	1808 (18.1)	1733 (17.3)	1776 (17.8)
2	8016 (80.2)	7991 (79.9)	7999 (80.0)	8068 (80.7)	8026 (80.3)
	Mean (standard-deviation)				
Interval time between the first child and the second child (years)	7.21 (5.05)	7.27 (5.07)	7.35 (5.18)	7.36 (5.12)	7.38 (5.18)

Notes: control group: without any intervention;

a: prospective father: high fertility intention, prospective mother: high fertility intention;

b: prospective father: high fertility intention, prospective mother: medium fertility intention;

c: prospective father: high fertility intention, prospective mother: low fertility intention;

d: prospective father: medium fertility intention, prospective mother: high fertility intention.

influence, which means the influencing relation could be positive, but also negative, namely, harmonious or stormy. If all three kinds of relations have above two situations, then the combinations will be $8 = 2 \times 2 \times 2$, corresponding to 8 experimental plans. Then, we run different experimental plans to simulate their effects on the generations of fertility behaviors and the number of new born children.

For each of experimental plans, we run 10,000 times (representing the simulation of 10,000 families), and in each run, the experimental variable is given value randomly according to a certain range as we designed. In each run and in each mutual influence in it, the current intention (at time t) of one family member is affected and determined by his/her own and other family members' former intentions (at time $t - 1$).

In the Table 1a and Table 1b, we present the outcomes of experimental program I. More than 40% of households have no child in the end of the two-year simulation. In

Table 1b. Experiment I's outcomes of families' fertility behaviors under adjusting couples' initial fertility intentions

Variable	Experimental outcome: numbers of families (percent of families)				
	e	f	g	h	i
No. of children (2 years)					
0	4160 (41.6)	4062 (40.6)	4192 (41.9)	4182 (41.8)	4066 (40.7)
1	3744 (37.4)	3827 (38.3)	3723 (37.2)	3692 (36.8)	3756 (37.6)
2	2096 (21.0)	2111 (21.1)	2085 (20.8)	2136 (21.4)	2178 (21.8)
No. of children (5 years)					
0	3447 (34.5)	3321 (33.2)	3440 (34.4)	3402 (34.0)	3340 (33.4)
1	3174 (31.7)	3238 (32.4)	3192 (31.9)	3143 (31.4)	3195 (32.0)
2	3379 (33.8)	3441 (34.4)	3368 (33.7)	3455 (34.6)	3465 (34.7)
No. of children (lifetime)					
0	200 (2.0)	205 (2.1)	196 (2.0)	205 (2.1)	197 (2.0)
1	1783 (17.8)	1841 (18.4)	1827 (18.3)	1744 (17.4)	1710 (17.1)
2	8017 (80.2)	7954 (79.5)	7977 (79.8)	8051 (80.5)	8093 (80.9)
	Mean (standard-deviation)				
Interval time between the first child and the second child (years)	7.40 (5.24)	7.28 (5.10)	7.40 (5.24)	7.35 (5.20)	7.34 (5.22)

Notes: *e*: prospective father: medium fertility intention, prospective mother: medium fertility intention;
f: prospective father: medium fertility intention, prospective mother: low fertility intention;
g: prospective father: low fertility intention, prospective mother: high fertility intention;
h: prospective father: low fertility intention, prospective mother: medium fertility intention;
i: prospective father: low fertility intention, prospective mother: low fertility intention.

the end of the five-year simulation, more than one third of households have two children. In addition, there is nearly no difference among control group and experimental plans' outcomes on the interval time between the first child and the second child.

In the Table 2a and Table 2b, we present the outcomes of experimental program II. Around 40% of households have no child in the end of the two-year simulation. In the end of the five-year simulation, more than one third of households have two children. Further, we can find that basically the more harmonious relations among family members, the higher ratio of families having the second child in the end of the lifetime simulation.

Table 2a. Experiment II's outcomes of families' fertility behaviors under adjusting relationships between family members

Variable	Experimental outcome: numbers of families (percent of families)			
	a	b	c	d
No. of children (2 years)				
0	3912 (39.1)	3887 (38.9)	3992 (39.9)	3949 (39.5)
1	3567 (35.7)	3684 (36.8)	3547 (35.5)	3688 (36.9)
2	2521 (25.2)	2429 (24.3)	2461 (24.6)	2363 (23.6)
No. of children (5 years)				
0	2920 (29.2)	2948 (29.5)	3056 (30.6)	3022 (30.2)
1	3054 (30.5)	3162 (31.6)	3061 (30.6)	3140 (31.4)
2	4026 (40.3)	3890 (38.9)	3883 (38.8)	3838 (38.4)
No. of children (lifetime)				
0	853 (8.5)	879 (8.8)	939 (9.4)	885 (8.9)
1	4994 (49.9)	5101 (51.0)	4950 (49.5)	4981 (49.8)
2	4153 (41.5)	4020 (40.2)	4111 (41.1)	4134 (41.3)
	Mean (standard-deviation)			
Interval time between the first child and the second child (years)	5.70 (4.57)	5.71 (4.55)	5.77 (4.43)	5.74 (4.56)

Notes: a: couples' relationships: great; relations between couples and their own parents: great; relations between couples and their parents-in-law: great;

b: couples' relationships: great; relations between couples and their own parents: great; relations between couples and their parents-in-law: normal;

c: couples' relationships: great; relations between couples and their own parents: normal; relations between couples and their parents-in-law: great;

d: couples' relationships: great; relations between couples and their own parents: normal; relations between couples and their parents-in-law: normal.

Table 2b. Experiment II's outcomes of families' fertility behaviors under adjusting relationships between family members

Variable	Experimental outcome: numbers of families (percent of families)			
	e	f	g	h
No. of children (2 years)				
0	4181 (41.8)	4197 (42.0)	4152 (41.5)	4137 (41.4)
1	3641 (36.4)	3697 (37.0)	3673 (36.7)	3759 (37.6)
2	2178 (21.8)	2106 (21.1)	2175 (21.8)	2104 (21.0)
No. of children (5 years)				
0	3364 (33.6)	3394 (33.9)	3389 (33.9)	3391 (33.9)
1	3141 (31.4)	3239 (32.4)	3109 (31.1)	3221 (32.2)
2	3495 (34.9)	3367 (33.7)	3502 (35.0)	3388 (33.9)
No. of children (lifetime)				
0	1090 (10.9)	1147 (11.5)	1219 (12.2)	1183 (11.8)
1	4884 (48.8)	4949 (49.5)	5027 (50.3)	5019 (50.2)
2	4026 (40.3)	3904 (39.0)	3754 (37.5)	3798 (38.0)
	Mean (standard-deviation)			
Interval time between the first child and the second child (years)	5.98 (4.78)	6.03 (4.68)	5.99 (4.66)	5.90 (4.58)

Notes: *e*: couples' relationships: normal; relations between couples and their own parents: great; relations between couples and their parents-in-law: great;

f: couples' relationships: normal; relations between couples and their own parents: great; relations between couples and their parents-in-law: normal;

g: couples' relationships: normal; relations between couples and their own parents: normal; relations between couples and their parents-in-law: great;

h: couples' relationships: normal; relations between couples and their own parents: normal; relations between couples and their parents-in-law: normal.

8   Discussion, Conclusion and Future Work

To model the mutual influence among agents' preferences (such as intentions) and the transformation from agents' preferences (intentions) to their behaviors in groups, we built a multi-agent system (MAS_{IITIB}) in the context of social networks. Further, we ran experiments and simulations for the study of the interaction of fertility intentions and the generation of fertility behaviors in families, from a micro perspective of individual interactions instead of conventional macro perspective of entire social system in Demography. Our agent-based model handles the complicated multiple influencing relationships (stronger or weaker, positive or negative) among individual members in families, and the transformation from family members' fertility intentions to prospective mothers' fertility behaviors with consideration of randomness. In our research, we designed a series of experimental programs, and monitored the dynamics of fertility intentions and behaviors, and particularly the number of new born children in the long-term. We hope that our results would contribute to discovering important factors that affect the fertility behavior and population growth under different policies (particularly under consideration of the policy changed from the one-child policy to the universal two-child policy in China) and help the government to adjust policies and strategies for the well-being of society. However, our work is not yet sufficient, as many other prospects remain. One aspect is about the advancements of the general model of influence among preferences and its transformation to behaviors:

- We express agents' preferences in a cardinal approach, assuming a normalized value as intention for each agent, and use the intention values of agents as probabilities for the generation of their behaviors to include randomness in the formulation of the transformation from preferences to behaviors. An ordinal approach to represent agents' preferences, particularly the preference ordering among multiple alternatives, could be tried in future work.
- We have not provided enough material to support the multi-agent system we designed to accurately characterize the dynamics of agents' preferences and behaviors due to mutual influence in groups in practice. In the future, the contributions of other disciplines such as logic and psychology, cognitive science and behavior science, particularly the empirical evidences in these fields, would be much valuable.
- We assumed a linear relation between agents' intentions and the generation of their behaviors: the higher an agent's intention is, the more possible it is to produce a behavior, and the return brought by the increase of intention is proportional. While in reality, there may be some thresholds of intentions for the generation of behaviors, which means the increase of intention within certain range may have very limited impacts, but once beyond a certain threshold, the return brought by the increase of intention will be much tremendous. The artificial neural network could be used to model the complicated psychical process under the transformation from intentions to behaviors of humans in future work, particularly making full use of all kinds of non-linear activation functions.

The other aspect is about the improvements of the agent-based modeling and simulation on fertility intention and behavior:

- We mainly modeled the impact of age on a prospective mother's fertility intention evolution, but have to admit that there are many other potential influential factors such as the income and expense, health condition, and balance with her life plan. From the economic (rational) perspective on fertility behavior, a utility-based approach may be recommended. In this approach, prospective parents may try to optimize the expected utility combing both assistance and care in the elderly period provided by their children and expense defrayed and energy consumed during the childhood of their children. Economists may encourage us to introduce to the agent-based model some economic variables.
- We assumed one usual household structure composed of a husband and a wife, both two (prospective) grandparents on each side, and several relatives and close friends on each side. But in real-world situations, there are more kinds of household structure, for instance, there are not four grandparents alive. More household

structures can be designed in future work, particularly according to the result of the census of population.

– We mainly discussed the interaction of fertility intentions and the generation of fertility behaviors in families under the impact of population policy transition. While the fertility intention and behavior are closely related to cultural traditions which are hard to be measured from a computational perspective. For instance, in China, the effectiveness of the implementation of population polices are quite different in different areas. Therefore, the fertility intention and behavior are affected not only by state interventions. The interplay of political, economic and cultural factors can be discussed in future work.

Moreover, with certain adjustments, there are a lot of potentials to use our model in studying the mutual-influence among preferences and the transformation from preferences to behaviors with various group settings.

Acknowledgements. This study is supported by a National Natural Science Foundation of China Grant (71804006), a National Natural Science Foundation of China and European Research Council Cooperation and Exchange Grant (7161101045), and a UKRI's Global Challenge Research Fund (ES/P011055/1). An earlier version of this study has been presented in the local proceedings of the 19th International Conference on Group Decision and Negotiation held at Loughborough, UK, thanks for the advices from reviewers and audiences.

References

1. Andrew, M., Meen, G.: Population structure and location choice: a study of London and South East England. Pap. Reg. Sci. **85**(3), 401–419 (2006)
2. Booth, H.: Demographic forecasting: 1980–2005 in review. Int. J. Forecast. **22**(3), 547–581 (2006)
3. Callick, R.: China relaxes its one-child policy. The Australian, 24 Jan 2007
4. Capuano, N., Chiclana, F., Fujita, H., Herrera-Viedma, E., Loia, V.: Fuzzy group decision making with incomplete information guided by social influence. IEEE Trans. Fuzzy Syst. **26**(3), 1704–1718 (2018)
5. Degroot, M.H.: Reaching a consensus. J. Am. Stat. Assoc. **69**(345), 118–121 (1974)
6. Demarzo, P.M., Vayanos, D., Zwiebel, J.: Persuasion bias, social influence, and unidimensional opinions. Q. J. Econ. **118**(3), 909–968 (2003)
7. Friedkin, N.E., Johnsen, E.C.: Social influence and opinions. J. Math. Soc. **15**(3–4), 193–206 (1990)
8. Friedkin, N.E., Johnsen, E.C.: Social positions in influence networks. Soc. Netw. **19**(3), 209–222 (1997)
9. Golub, B., Jackson, M.O.: Naive learning in social networks and the wisdom of crowds. Am. Econ. J. Microecon. **2**(1), 112–149 (2010)
10. Gu, B., Wang, F., Guo, Z., Zhang, E.: China's local and national fertility policies at the end of the twentieth century. Popul. Dev. Rev. **33**(1), 129–148 (2007)
11. Guo, Z.: Changes in family households in China in 1990s. Paper presented at The Academic Conference on the 2000 Population Census in China, 28–31 March, Beijing (2003). (in Chinese)

12. Grabisch, M., Rusinowska, A.: A model of influence based on aggregation function. Math. Soc. Sci. **66**(3), 316–330 (2013)

13. Grabisch, M., Rusinowska, A.: A model of influence in a social network. Theory Decis. **69** (1), 69–96 (2010). https://doi.org/10.1007/s11238-008-9109-z

14. Grabisch, M., Rusinowska, A.: A model of influence with an ordered set of possible actions. Theory Decis. **69**(4), 635–656 (2010). https://doi.org/10.1007/s11238-009-9150-6

15. Grabisch, M., Rusinowska, A.: Influence functions, followers and command games. Games Econ. Behav. **72**(1), 123–138 (2011)

16. Grabisch, M., Rusinowska, A.: Measuring influence in command games. Soc. Choice Welfare **33**(2), 177–209 (2009). https://doi.org/10.1007/s00355-008-0350-8

17. Grandi, U., Lorini, E., Perrussel, L.: Propositional opinion diffusion. In: Proceedings of the 14th International Conference on Autonomous Agents and Multiagent Systems, pp. 989–997 (2015)

18. Keilman, N., Pham, D.Q.: Predictive intervals for age-specific fertility. Eur. J. Popul. **16**(1), 41–65 (2000). https://doi.org/10.1023/A:1006385413134

19. Luo, H., Meng, Q.: Multi-agent simulation of SC reform and national game. World Econ. Polit. **6**, 136–155 (2013). (in Chinese)

20. Luo, H.: How to address multiple sources of influence in group decision-making? In: Morais, D.C., Carreras, A., de Almeida, A.T., Vetschera, R. (eds.) GDN 2019. LNBIP, vol. 351, pp. 17–32. Springer, Cham (2019). https://doi.org/10.1007/978-3-030-21711-2_2

21. Luo, H.: Individual, coalitional and structural influence in group decision-making. In: Torra, V., Narukawa, Y., Pasi, G., Viviani, M. (eds.) MDAI 2019. LNCS (LNAI), vol. 11676, pp. 77–91. Springer, Cham (2019). https://doi.org/10.1007/978-3-030-26773-5_7

22. Lutz, W.: Scenario analysis in population projection. International Institute for Applied Systems Analysis [IIASA], Laxenburg (1995)

23. Maran, A., Maudet, N., Pini, M.S., Rossi, F., Venable, K.B.: A framework for aggregating influenced CP-nets and its resistance to bribery. In: Proceedings of the Twenty-Seventh AAAI Conference on Artificial Intelligence, pp. 668–674 (2013)

24. Maudet, N., Pini, M.S., Venable, K.B., Rossi, F.: Influence and aggregation of preferences over combinatorial domains. In: Proceedings of the 11th International Conference on Autonomous Agents and Multiagent Systems, pp. 1313–1314 (2012)

25. National Health and Family Planning Commission of the PRC. Q&A about the New Two-Child Policy. http://en.nhfpc.gov.cn/2015-11/06/c_45715.htm. Accessed 01 Nov 2018

26. Prez, L.G., Mata, F., Chiclana, F., Gang, K., Herrera-Viedma, E.: Modelling influence in group decision making. Soft. Comput. **20**(4), 1653–1665 (2016). https://doi.org/10.1007/s00500-015-2002-0

27. Salehi-Abari, A., Boutilier, C.: Empathetic social choice on social networks. In: Proceedings of the 13th International Conference on Autonomous Agents and Multiagent Systems, pp. 693–700 (2014)

28. Wang, F., Cai, Y., Gu, B.: Population, policy, and politics: how will history judge China's one-child policy? Popul. Dev. Rev. **38**, 115–129 (2013)

29. Wang, Z., Dang, Y., Wang, Y.: A new grey Verhulst model and its application. In: Proceedings of the IEEE International Conference on Grey Systems and Intelligent Services (2007)

30. Zeng, Y., Therese, H.: The effects of China's universal two-child policy. The Lancet **388** (10054), 1930–1938 (2016)

31. Zhai, Z., Li, L., Chen, J.: Accumulated couples and extra births under the universal two-child policy. Popul. Res. **4**, 35–51 (2016). (in Chinese)

32. Zheng, B.: Population ageing and the impacts of the universal two-child policy on China's socio-economy. Econ. Polit. Stud. **4**(4), 434–453 (2016)

Manipulability of Majoritarian Procedures in Two-Dimensional Downsian Model

Daniel Karabekyan[1](✉) and Vyacheslav Yakuba[1,2]

[1] National Research University Higher School of Economics,
20, Myasnitskaya Street, Moscow 101000, Russia
dkarabekyan@hse.ru
[2] Institute of Control Sciences, Russian Academy of Sciences,
65, Profsoyuznaya Street, Moscow 117997, Russia

Abstract. For the two-dimensional Downsian model the degree of manipulability of 16 known aggregation procedures, based on the majority relation, is evaluated using the Nitzan-Kelly index. Extended preferences for multi-valued choices are used to evaluate the fact of manipulation. Individual manipulability of agents is considered, when manipulating agent moves its ideal point over the plane. The range of possible manipulating positions of the agents is restricted to some rectangle on the two-dimensional coordinate space, within the feasible area of positions of alternatives and agents. The preferences of agents are assumed to be linear orders, constructed by the proximity of the alternatives to the agents, ordered according to Euclidean distance. The computer calculations, using Monte-Carlo simulations has been performed for 3, 4, and 5 alternatives and for even number of agents from 4 to 20. 100 thousands profiles were generated for each number of alternatives – number of agents case. The results of the simulations show that there are groups of procedures with relatively low degree of manipulability for all of the considered multiple-choice extensions.

Keywords: Manipulability · Majoritarian choice procedures · Spatial Downsian model

1 Manipulability Model

The spatial model of voting considered in [9, 10], and in many other works (see e.g. [11]), gives the new insight to understanding voting procedures. We consider here the new model of manipulation in the spatial model. The manipulability in the IC, IAC models is considered in [4], and the manipulability in one-dimensional Downsian model is evaluated in [2]. For the spatial model, in [12] it is inferred, that least manipulable among the positional methods with any number of alternatives is the Borda rule, while the plurality rule is one of the most manipulable of the positional methods.

In the framework of the two-dimensional Downsian model [6] the profiles are generated, representing a set of preferences of agents. There are m alternatives, (m = 3, 4, or 5) and n ideal points of agents (n > 3) on the rectangle in the planar space. For each agent, her preferences are constructed in the following way. The Euclidean

D. C. Morais et al. (Eds.): GDN 2020, LNBIP 388, pp. 120–132, 2020.
https://doi.org/10.1007/978-3-030-48641-9_9

distance from the ideal point of the particular agent to each of the alternatives is calculated. The alternatives are ordered by proximity to the ideal point, the closest alternative is put on the first place, the next on the second place, etc., the farthest alternative is put on the last place.

After positions of agents and alternatives are determined, agents can manipulate by reporting insincere preferences. Agents can shift their ideal position in such a way that provides them better result of collective choice. There are several different methods to do so. First, the agent can only move its position inside the feasible area, in this case, a rectangle. Second, the agent can move throughout the entire space. And third, an agent can reveal any preference, regardless of whether it is acceptable given the current location of the alternatives. In the second version of manipulation, the agent's insincere position, in the case of going beyond the boundaries of the limited space, may not be permissible from the substantive considerations in terms of interpretations of the axes in space. In the third version of manipulation, the presentation of unacceptable preferences can be unambiguously recognized by other participants in the vote as manipulation. Thus, in the evaluations, the first version of manipulation is considered. Tie-breaking situations and degenerated cases are not considered.

In the two-dimensional Downsian model, for example, for 3 alternatives, for the non-degenerated case, in the entire space, all orderings of the alternatives are feasible. However, if the agent for the purpose of manipulation is forced to move far beyond the boundaries of the considered limited space, usually a rectangle, then such orderings are not considered as the possible ones.

Let us consider the example of manipulation for 3 alternatives, 4 agents, for the Leximin extension of preferences for the Minimal dominant set procedure. Let the alternatives and agents are located on the planar space as shown on the left image of Fig. 1. Note that negative positions of the Y axis are also possible.

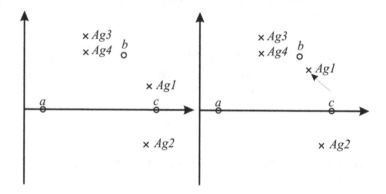

Fig. 1. Example of the manipulation in two-dimensional space.

Let agent Ag1 manipulate, shifting her position towards alternative b as shown on the right image of Fig. 1.

For each agent, the ordering of alternatives is built by proximity to the ideal point of the agent. For example, for Ag1, the ordering is as follows: c > b > a.

The profile consisting of the preferences of 4 agents for 3 alternatives is presented in Table 1.

Table 1. The profile for 3 alternatives, 4 agents for sincere positions on Fig. 1

Ag1	Ag2	Ag3	Ag4
c	c	b	b
b	b	a	a
a	a	c	c

The majority relation matrix has the following form (see definition of the majority relation below).

Table 2. Majority matrix for the profile in Table 1.

	a	b	c
a	0	0	0
b	1	0	0
c	0	0	0

In this matrix in Table 2 the number "1" means that the alternative b dominates the alternative a. "0" means that the pair of alternatives does not dominate each other.

The choice according to the Minimal dominant set is {a, b, c} (see the definitions of the aggregation procedures below). If the ordering of the agent Ag1 according to sincere preferences was c > b > a, then after manipulating the agent reports the following ordering of alternatives b > c > a. The profile after manipulation of the agent Ag1 has the form (Table 3):

Table 3. Example profile after manipulation.

Ag1	Ag2	Ag3	Ag4
b	c	b	b
c	b	a	a
a	a	c	c

The corresponding majority relation matrix has the form

Table 4. Majority matrix after manipulation.

	a	b	c
a	0	0	0
b	1	0	1
c	0	0	0

In the matrix in Table 4 the number "1" means that the alternative in the row dominates the alternative in the column, that is b > a and b > c. The number "0" means that there is no domination between the pair of the alternatives.

The choice made by the Minimal dominant set after manipulation by the agent Ag1 is {b}.

In accordance with the Leximin extension, for the Ag1 agent, the choice resulting from manipulation is more preferable than that with sincere preferences. Thus, the profile shown in Fig. 1. is manipulable for the Minimum Dominant Set procedure, for the Leximin extension.

Manipulability is evaluated for the following aggregation procedures based on the majoritarian relation.

Minimal Dominant Set (MDS), Minimal Undominated Set (MUS), Minimal Weakly Stable Set (MWSS), Fishburn's rule (F), Uncovered Set I, II (US1, US2), Richelson's rule (R), Copeland's rule I, II, III (C1, C2, C3), k-stable set, k = 1, 2, 3, (kSS), k-stable set II, k = 1, 2, 3, (kSSII).

The description of the procedures, except for the k-stable set rule, is given in [5]. Different versions of the k-stable set were introduced in [3]. In [1], extended preferences were introduced, in which many alternatives are ordered based on the single-valued preferences of the agents. To evaluate the degree of manipulability of aggregation procedures, the Nitzan – Kelly index is used, which was introduced in [7, 8]. The index is defined as

$$NK = \frac{d_0}{d_{total}},$$

where d_0 - number of manipulable profiles, d_{total} - total number of profiles.

2   The Aggregation Procedures

We present definitions of the choice procedures, for which the degree of manipulability is evaluated.

The majority relation μ is defined as follows. For some alternatives x and y from the set of alternatives A it is said, that alternative x dominates the alternative y via majority relation μ ($x\mu y$) if the following condition holds. The number of agents for which x better, than y exceeds the number of agents with the opposite preferences (y is better, than x). The Upper counter set of an alternative x in the relation P is the set $D(x)$ such that $D(x) = \{y \in A \mid yPx\}$. The Lower counter set of x in the relation P is the set $L(x)$ such that $L(x) = \{y \in A \mid xPy\}$.

Now we list the considered rules and discuss the choice of the sample profile. We follow the definitions and notations of [3, 5].

1. (MDS) Minimal dominant set. The set Q is called dominant set if any alternative in Q dominates each alternative outside Q via majority relation μ. The dominant set Q is said to be the minimal if no proper subset is a dominant set. The choice is equal to Q. If there are several of them, then the union is taken.

2. (MUS) Minimal Undominated Set. The set Q is called undominated set if no alternative outside Q dominates any alternative inside Q via majority relation μ. The undominated set Q is a minimal one, if no proper subset is an undominated set. The choice is equal to Q. If the set is not unique, then the union of such sets is taken.

3. (MWSS) Minimal Weakly Stable Set. The alternative x belongs to the weakly stable set Q if the following condition holds. If there exists an alternative $y \in A \backslash Q$ which dominates x via majority relation μ, then there exists some alternative $z \in Q$, which dominates y, i.e., $z\mu y$. A set is called the minimal weakly stable if it does not contain proper weakly stable subsets. The choice consists of the alternatives of the set Q. If the set is not unique, then the union of all such sets is taken.

4. (F) Fishburn's Rule. The binary relation γ is defined as follows $x\gamma y \Leftrightarrow D(x) \subset D(y)$, where $D(x)$ is the upper contour set of the alternative x in the majority relation μ. The choice contains the undominated alternatives via γ.

5. (US1) Uncovered Set I. We construct a binary relation δ in the following way $x\delta y \Leftrightarrow L(x) \supset L(y)$, where $L(x)$ is the lower contour sets of the alternative x in the majority relation μ. The undominated alternatives on δ are chosen.

6. (US2) Uncovered Set II. The alternative x B-dominates some alternative y (xBy) if $x\mu y$ and $D(x) \subseteq D(y)$, where $D(x)$ is the upper contour set of x in μ. The choice consists of B-undominated alternatives.

7. (R) Richelson's Rule. For the majority relation μ upper and lower contour sets ($D(x)$ and $L(x)$) for each alternative $x \in A$ in the relation μ are constructed. Then the binary relation σ is defined as follows:

$$x\sigma y \Leftrightarrow [L(x) \supseteq L(y) \wedge D(x) \subseteq D(y) \wedge ([L(x) \supset L(y)] \vee [D(x) \subset D(y)])]$$

The choice consists of the alternatives undominated via σ.

8. (C1) Copeland's rule I. For each alternative x the value function $u(x)$ is constructed as the difference of the cardinalities of lower and upper contour sets of the alternative in majority relation μ. The choice consists of the alternatives with maximum value of u.

9. (C2) Copeland's rule II. Function $u(x)$ is equal to the cardinality of the lower contour set of alternative x in majority relation μ. The alternatives with maximum of u are chosen.

10. (C3) Copeland's rule III. Function $u(x)$ is equal to the cardinality of the upper contour set of the alternative x in majority relation μ. The alternatives with minimum of u are chosen.

11. (kSS) k-stable set.
    The alternative x belongs to the k-stable set for $k = 1, 2, 3$, if one of following two conditions hold:
    a) the alternative x is undominated by alternatives outside the k-stable set via majority relation μ, or
    b) if $\exists y \notin kSS : y\mu x$ then $\exists z_1, \ldots, z_k, z_1 \in kSS$ and $z_1 \mu z_2 \mu \ldots z_k \mu y$. In general, $k = 1, 2,$ or 3. The k-stable set is minimal, if no proper subsets of it are k-stable. If there are several different k-stable sets, then the union is taken.

12. (kSSII) k-stable set II.
    The k-stable set II consists of the alternatives, that are not dominated by alternatives outside the set. And for each alternative outside the set, there is an alternative

in the set, which dominates it via majority relation in not more than k steps. k steps here means that $\exists z_1, \ldots, z_k$, $z_1 \in kSS$ and $z_1 \mu z_2 \mu \ldots z_k \mu y$.

3 Extended Preferences

For the case of multiple choice, the sets of alternatives also should be compared as an extension to the agents' preferences. In total, as introduced and discussed in [1], there are 4 extensions for 3 alternatives, 10 extensions for 4 alternatives, and 12 extensions for 5 alternatives. In the following formulae the extensions differ in underlined parts.

For 3 alternatives, the extended preferences look as follows, (agent's single-valued ordering is assumed to be a > b > c):

Leximin (3 alts): $\{a\} \succ \{a,b\} \succ \underline{\{b\}} \succ \{a,c\} \succ \{a,b,c\} \succ \{b,c\} \succ \{c\}$

Leximax (3 alts): $\{a\} \succ \{a,b\} \succ \underline{\{a,b,c\}} \succ \{a,c\} \succ \{b\} \succ \{b,c\} \succ \{c\}$

PWorst (3 alts): $\{a\} \succ \{a,b\} \succ \underline{\{b\}} \succ \{a,b,c\} \succ \{a,c\} \succ \{b,c\} \succ \{c\}$

PBest (3 alts): $\{a\} \succ \{a,b\} \succ \underline{\{a,c\}} \succ \{a,b,c\} \succ \{b\} \succ \{b,c\} \succ \{c\}$

For 4 alternatives let us preset one out of ten extensions (single-valued ordering of the agent is assumed to be a > b > c > d).

PWorst (4 alts):

$$\{a\} \succ \{a,b\} \succ \{b\} \succ \{a,b,c\} \succ \{a,c\} \succ \{b,c\} \succ \{a,b,c,d\}$$

$$\succ \{a,b,d\} \succ \{a,c,d\} \succ \{b,c,d\} \succ \{a,d\} \succ \{b,d\} \succ \{c,d\} \succ \{d\}$$

For 5 alternatives let us preset only one out of twelve extensions (single-valued ordering of the agent is assumed to be a > b > c > d > e).

Rank increasing power Leximin (5 alts):

$$\{a\} \succ \{a,b\} \succ \underline{\{b\}} \succ \{a,c\} \succ \{a,b,c\} \succ \{a,b,d\} \succ$$

$$\underline{\{b,c\}} \succ \{a,d\} \succ \{a,b,c,d\} \succ \{a,c,d\} \succ \{a,b,e\} \succ \{a,b,c,e\} \succ$$

$$\{c\} \succ \underline{\{b,d\}} \succ \{a,e\} \succ \{b,c,d\} \succ \{a,c,e\} \succ$$

$$\{a,b,d,e\} \succ \{a,b,c,d,e\} \succ$$

$$\{a,c,d,e\} \succ \underline{\{b,c,e\}} \succ \{a,d,e\} \succ$$

$$\underline{\{c,d\}} \succ \{b,e\} \succ \{b,c,d,e\} \succ \{b,d,e\} \succ$$

$$\underline{\{d\}} \succ \{c,e\} \succ \{c,d,e\} \succ \{d,e\} \succ \{e\}$$

The preference extensions are defined axiomatically. Formal definitions of all extended preferences and detailed discussion is presented in [1].

4 The Scheme of Calculation

The model, in general, is based on the following intuition. Let us place alternatives and agents on the two-dimensional space. Let us also construct a profile of preferences, in which each agent's ordering is induced by the distances of the alternatives to this agent. Then we move the position of each agent individually within some area, not too far from the positions of the alternatives and other agents. The changes in the profiles yield different choices, obtained via the aggregation procedures. If the result of such shift of the position is better for the agent, then the profile is manipulable. The goal is to evaluate the share of such manipulable profiles for each aggregation procedure considered.

The calculation is performed in the following way. On the two-dimensional coordinate space, the rectangle, centered at the point [0; 0], with side X to side Y ratio 1:2 is defined. The restricted rectangle area can be formally defined as set of points with coordinates $0 \le x \le 1$ and $-1 \le y \le 1$.

There are m alternatives (m = 3, 4, 5) and n agents (n = 4–20), even number, generated randomly with X coordinates from 0 to 1 and Y from -1 to 1. Without loss of generality we can assume that in any profile there are two alternatives that are farthest from each other, for example, a and e in the case of 5 alternatives, are placed in fixed positions: a in [0,0], and e in [1,0] and all alternatives are sorted alphabetically along the X axe. For each agent, the ordering of sincere preferences is constructed according to the distance from the agent's ideal point to alternatives. For the generated profile, the choice is calculated for each of the considered aggregation procedures.

Individual manipulation by agents is considered. The evaluations are performed in the following way. For each agent individually, on the grid of L steps, the agent's manipulating point with the coordinates (lx, ly) for lx and ly with a step of 1/L is positioned. For the presented results, L = 100. The agent's insincere preferences are constructed, as if the manipulating point was the ideal point of the agent. Alternatives are ordered by proximity to this point. For a profile constructed in such a way with insincere preferences of the manipulating agent, the choice is evaluated according to all considered aggregation procedures.

For a manipulating agent, for each choice procedure, the choices obtained by sincere preferences and by insincere preferences are compared using all considered preferences extensions. (The results are presented for Leximin, Leximax, PWorst, and PBest extensions for 3 alternatives, for PWorst extension for 4 alternatives, and for Rank increasing power leximin for 5 alternatives.) If the collective choice obtained on the profile, which consists of insincere preferences of the manipulating agent is better for the manipulating agent than a choice obtained from the sincere preferences, then manipulation is considered to be a successful one for a given profile, a given rule and the type of extended preferences.

For each rule and extended preferences the number of manipulable profiles is calculated, and the Nitzan-Kelly manipulability index is calculated as the ratio of the number of manipulated profiles to the total number of generated profiles.

5 Calculation Results

For even number of agents, the procedures MDS, F, US1, R, C1, C2 have high degree of manipulability for most extensions in the cases for 3, 4, and 5 alternatives.

We present the results of the evaluation of the degree of individual manipulability for 3, 4, and 5 alternatives in two-dimensional space for even number of agents from 4 to 20.

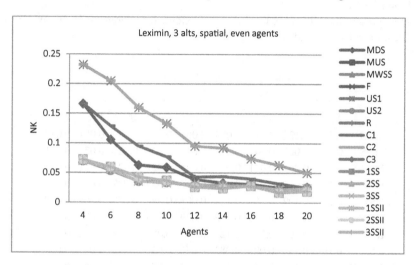

Fig. 2. Manipulability of aggregation procedures for 3 alternatives, Leximin extension, even number of agents.

Table 5. NK index, 3 alts, Leximin.

	Agents								
	4	6	8	10	12	14	16	18	20
MDS	0.166	0.106	0.063	0.059	0.039	0.033	0.031	0.025	0.026
MUS	0.072	0.06	0.043	0.037	0.026	0.024	0.028	0.017	0.019
MWSS	0.072	0.06	0.043	0.037	0.026	0.024	0.028	0.017	0.019
F	0.072	0.06	0.043	0.037	0.026	0.024	0.028	0.017	0.019
US1	0.232	0.205	0.16	0.133	0.095	0.092	0.074	0.063	0.05
US2	0.07	0.054	0.036	0.034	0.028	0.03	0.028	0.022	0.025
R	0.232	0.205	0.16	0.133	0.095	0.092	0.074	0.063	0.05
C1	0.166	0.129	0.095	0.077	0.043	0.044	0.04	0.032	0.025
C2	0.232	0.205	0.16	0.133	0.095	0.092	0.074	0.063	0.05
C3	0.072	0.06	0.043	0.037	0.026	0.024	0.028	0.017	0.019
1SS	0.072	0.06	0.043	0.037	0.026	0.024	0.028	0.017	0.019
2SS	0.072	0.06	0.043	0.037	0.026	0.024	0.028	0.017	0.019
3SS	0.072	0.06	0.043	0.037	0.026	0.024	0.028	0.017	0.019
1SSII	0.07	0.056	0.036	0.033	0.031	0.028	0.028	0.022	0.024
2SSII	0.072	0.06	0.043	0.037	0.026	0.024	0.028	0.017	0.019
3SSII	0.072	0.06	0.043	0.037	0.026	0.024	0.028	0.017	0.019

For example, for 3 alternatives for the Leximin extension, it can be seen from the Fig. 2. and Table 5, that there are 3 groups of procedures with more or less same level of manipulability. The high-manipulable group consists of US1, R, and C2. There are MDS and C1 procedures, which have middle level of manipulability for low number of agents and then their level of manipulability decreases. The third group with small level of manipulability consists of MUS, MWSS, F, US2, C3, 1SS, 2SS, 3SS, and 1SSII, 2SSII, and 3SSII.

Let us present the NK index for the rest of extended preferences, Leximax, PWorst, and PBest (Tables 6, 7 and 8).

Table 6. NK index, 3 alts, Leximax.

	Agents								
	4	6	8	10	12	14	16	18	20
MDS	0.114	0.122	0.11	0.094	0.081	0.062	0.05	0.051	0.037
MUS	0.1	0.067	0.041	0.033	0.025	0.027	0.027	0.021	0.021
MWSS	0.1	0.067	0.041	0.033	0.025	0.027	0.027	0.021	0.021
F	0.1	0.067	0.041	0.033	0.025	0.027	0.027	0.021	0.021
US1	0.24	0.197	0.154	0.129	0.091	0.088	0.073	0.062	0.048
US2	0.1	0.069	0.044	0.038	0.029	0.034	0.028	0.023	0.024
R	0.24	0.197	0.154	0.129	0.091	0.088	0.073	0.062	0.048
C1	0.19	0.144	0.098	0.082	0.046	0.05	0.047	0.041	0.03
C2	0.24	0.197	0.154	0.129	0.091	0.088	0.073	0.062	0.048
C3	0.1	0.067	0.041	0.033	0.025	0.027	0.027	0.021	0.021
1SS	0.1	0.067	0.041	0.033	0.025	0.027	0.027	0.021	0.021
2SS	0.1	0.067	0.041	0.033	0.025	0.027	0.027	0.021	0.021
3SS	0.1	0.067	0.041	0.033	0.025	0.027	0.027	0.021	0.021
1SSII	0.194	0.14	0.1	0.087	0.069	0.061	0.053	0.035	0.04
2SSII	0.1	0.067	0.041	0.033	0.025	0.027	0.027	0.021	0.021
3SSII	0.1	0.067	0.041	0.033	0.025	0.027	0.027	0.021	0.021

Table 7. NK index, 3 alts, PWorst.

	Agents								
	4	6	8	10	12	14	16	18	20
MDS	0.185	0.131	0.089	0.085	0.055	0.044	0.045	0.041	0.034
MUS	0.1	0.067	0.044	0.034	0.023	0.025	0.027	0.017	0.018
MWSS	0.1	0.067	0.044	0.034	0.023	0.025	0.027	0.017	0.018
F	0.1	0.067	0.044	0.034	0.023	0.025	0.027	0.017	0.018
US1	0.232	0.205	0.157	0.132	0.095	0.088	0.074	0.062	0.049
US2	0.1	0.069	0.044	0.038	0.029	0.034	0.028	0.023	0.024
R	0.232	0.205	0.157	0.132	0.095	0.088	0.074	0.062	0.049
C1	0.166	0.129	0.095	0.077	0.043	0.044	0.04	0.032	0.025

(continued)

Table 7. (*continued*)

	Agents								
	4	6	8	10	12	14	16	18	20
C2	0.232	0.205	0.157	0.132	0.095	0.088	0.074	0.062	0.049
C3	0.1	0.067	0.044	0.034	0.023	0.025	0.027	0.017	0.018
1SS	0.1	0.067	0.044	0.034	0.023	0.025	0.027	0.017	0.018
2SS	0.1	0.067	0.044	0.034	0.023	0.025	0.027	0.017	0.018
3SS	0.1	0.067	0.044	0.034	0.023	0.025	0.027	0.017	0.018
1SSII	0.158	0.102	0.062	0.051	0.036	0.032	0.031	0.025	0.025
2SSII	0.1	0.067	0.044	0.034	0.023	0.025	0.027	0.017	0.018
3SSII	0.1	0.067	0.044	0.034	0.023	0.025	0.027	0.017	0.018

Table 8. NK index, 3 alts, PBest.

	Agents								
	4	6	8	10	12	14	16	18	20
MDS	0.247	0.186	0.132	0.113	0.092	0.072	0.058	0.048	0.048
MUS	0.072	0.06	0.04	0.036	0.028	0.026	0.028	0.021	0.022
MWSS	0.072	0.06	0.04	0.036	0.028	0.026	0.028	0.021	0.022
F	0.072	0.06	0.04	0.036	0.028	0.026	0.028	0.021	0.022
US1	0.24	0.197	0.157	0.13	0.091	0.092	0.073	0.063	0.049
US2	0.07	0.054	0.036	0.034	0.028	0.03	0.028	0.022	0.025
R	0.24	0.197	0.157	0.13	0.091	0.092	0.073	0.063	0.049
C1	0.19	0.144	0.098	0.082	0.046	0.05	0.047	0.041	0.03
C2	0.24	0.197	0.157	0.13	0.091	0.092	0.073	0.063	0.049
C3	0.072	0.06	0.04	0.036	0.028	0.026	0.028	0.021	0.022
1SS	0.072	0.06	0.04	0.036	0.028	0.026	0.028	0.021	0.022
2SS	0.072	0.06	0.04	0.036	0.028	0.026	0.028	0.021	0.022
3SS	0.072	0.06	0.04	0.036	0.028	0.026	0.028	0.021	0.022
1SSII	0.106	0.096	0.077	0.07	0.065	0.059	0.052	0.037	0.041
2SSII	0.072	0.06	0.04	0.036	0.028	0.026	0.028	0.021	0.022
3SSII	0.072	0.06	0.04	0.036	0.028	0.026	0.028	0.021	0.022

For 4 alternatives, the procedures MUS, MWSS, F, C3, 1SS, 2SS, 3SS, 2SSII, 3SSII have low manipulability for all extended preference types. For some of the extensions, the manipulability of US2 and 1SSII procedures is close to the minimal one (Table 9, Fig. 3).

Table 9. NK index, 4 alts, PWorst.

	Agents								
	4	6	8	10	12	14	16	18	20
MDS	0.232	0.164	0.107	0.089	0.053	0.041	0.036	0.034	0.032
MUS	0.153	0.089	0.059	0.041	0.029	0.021	0.017	0.014	0.017
MWSS	0.153	0.089	0.059	0.041	0.029	0.021	0.017	0.014	0.017
F	0.153	0.089	0.059	0.041	0.029	0.021	0.017	0.014	0.017
US1	0.379	0.251	0.223	0.165	0.124	0.09	0.074	0.053	0.059
US2	0.155	0.11	0.068	0.05	0.04	0.027	0.023	0.022	0.021
R	0.379	0.253	0.223	0.165	0.124	0.091	0.074	0.053	0.06
C1	0.271	0.165	0.124	0.098	0.072	0.046	0.042	0.031	0.032
C2	0.353	0.239	0.205	0.154	0.113	0.082	0.066	0.049	0.055
C3	0.153	0.089	0.059	0.041	0.029	0.021	0.018	0.016	0.017
1SS	0.153	0.089	0.059	0.041	0.029	0.021	0.017	0.014	0.017
2SS	0.153	0.089	0.059	0.041	0.029	0.021	0.017	0.013	0.018
3SS	0.153	0.089	0.059	0.041	0.029	0.021	0.017	0.013	0.018
1SSII	0.192	0.131	0.081	0.057	0.043	0.028	0.023	0.022	0.021
2SSII	0.153	0.09	0.059	0.041	0.029	0.021	0.017	0.014	0.017
3SSII	0.153	0.089	0.059	0.041	0.029	0.021	0.017	0.014	0.017

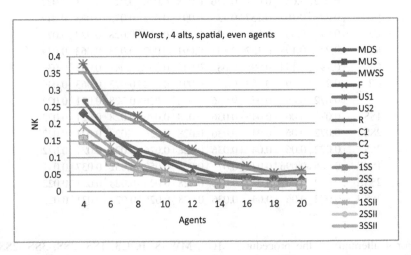

Fig. 3. Manipulability of choice procedures for 4 alternatives, PWorst extension, even number of agents.

For 5 alternatives, the same procedures: MUS, MWSS, F, US2, C3, 1SS, 2SS, 3SS, 1SSII, 2SSII, 3SSII show low level of manipulability (Table 10, Fig. 4).

Table 10. NK index, 5 alts, Rank increasing power leximin.

	Agents								
	4	6	8	10	12	14	16	18	20
MDS	0.385	0.31	0.222	0.178	0.147	0.126	0.106	0.084	0.086
MUS	0.215	0.156	0.102	0.085	0.073	0.055	0.052	0.044	0.045
MWSS	0.215	0.156	0.102	0.085	0.073	0.055	0.051	0.044	0.045
F	0.215	0.156	0.102	0.085	0.073	0.055	0.052	0.044	0.044
US1	0.514	0.444	0.34	0.267	0.221	0.202	0.182	0.138	0.123
US2	0.246	0.18	0.135	0.11	0.092	0.069	0.069	0.053	0.051
R	0.528	0.449	0.344	0.268	0.221	0.203	0.182	0.138	0.123
C1	0.352	0.292	0.223	0.155	0.13	0.107	0.099	0.069	0.069
C2	0.449	0.396	0.321	0.253	0.209	0.19	0.176	0.133	0.118
C3	0.215	0.156	0.103	0.086	0.073	0.055	0.052	0.044	0.046
1SS	0.215	0.156	0.102	0.085	0.073	0.055	0.051	0.044	0.045
2SS	0.215	0.156	0.101	0.086	0.072	0.054	0.051	0.043	0.043
3SS	0.215	0.156	0.101	0.086	0.073	0.055	0.051	0.043	0.044
1SSII	0.283	0.232	0.172	0.138	0.116	0.088	0.078	0.072	0.068
2SSII	0.217	0.157	0.102	0.087	0.073	0.056	0.051	0.044	0.044
3SSII	0.215	0.156	0.102	0.085	0.073	0.055	0.052	0.044	0.045

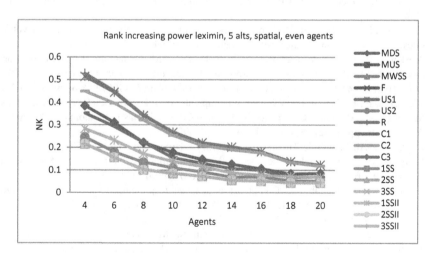

Fig. 4. Manipulability of aggregation procedures for 5 alternatives, Rank increasing power leximin, even number of agents.

6 Conclusion

We have considered a new model of manipulability of main majoritarian procedures in the Downsian model with extended preferences and show that almost all of them in this model have rather low level of manipulability. It is interesting to note that the following procedures have the level of manipulability close to minimum: Minimal Undominated Set, Minimal Weakly Stable Set, Fishburn's rule, Uncovered Set II, k-stable set, and k-stable set II, for k = 1, 2, 3.

Next interesting question is how manipulable are different scoring procedures. In the current model the rectangle area of feasible points of alternatives and agents is restricted in y coordinates by $-1 \leq y \leq 1$. It would be interesting to evaluate the sensitivity of the manipulability level depending on the size of the area.

Acknowledgements. The paper was prepared within the framework of the HSE University Basic Research Program and funded by the Russian Academic Excellence Project '5-100'. We thank the International Center of Decision Choice and Analysis (DeCAn Center) of the National Research University Higher School of Economics. The work is partially supported by RFBR grant #18-01-00804a. We are grateful to the anonymous referee for valuable comments.

References

1. Aleskerov, F., Karabekyan, D., Sanver, M.R., Yakuba, V.: An individual manipulability of positional voting rules. SERIEs 2(4), 431–446 (2011). https://doi.org/10.1007/s13209-011-0050-y
2. Aleskerov, F.T., Karabekyan, D., Ivanov, A., Yakuba, V.I.: Individual manipulability of majoritarian rules for one-dimensional preferences. Procedia Comput. Sci. 139, 212–220 (2018). 6th International Conference on Information Technology and Quantitative Management
3. Aleskerov, F., Karabekyan, D., Sanver, M.R., Yakuba, V.: On the manipulability of voting rules: the case of 4 and 5 alternatives. Math. Soc. Sci. 64(1), 67–73 (2012)
4. Aleskerov, F., Ivanov, A., Karabekyan, D., Yakuba, V.: Manipulability of majority relation-based collective decision rules. In: Czarnowski, I., Howlett, R.J., Jain, L.C. (eds.) IDT 2017. SIST, vol. 72, pp. 82–91. Springer, Cham (2018). https://doi.org/10.1007/978-3-319-59421-7_8
5. Aleskerov, F., Kurbanov, E.: A degree of manipulability of known social choice procedures. In: Alkan, A., Aliprantis, Ch., Yannelis, N. (eds.) Current Trends in Economics: Theory and Applications, pp. 13–27. Springer, Heidelberg (1999)
6. Downs, A.: An Economic Theory of Democracy. Harper, New York (1957)
7. Kelly, J.: Almost all social choice rules are highly manipulable, but few aren't. Soc. Choice Welfare 10, 161–175 (1993). https://doi.org/10.1007/BF00183344
8. Nitzan, S.: The vulnerability of point-voting schemes to preference variation and strategic manipulation. Public Choice 47, 349–370 (1985). https://doi.org/10.1007/BF00127531
9. McKelvey, R.D., Ordeshook, P.C.: A decade of experimental research on spatial models of elections and committees. In: Enelow, J.M., Hinich, M.J. (eds.) Advances in the Spatial Theory of Voting, pp. 99–144. Cambridge University Press, Cambridge (1990)
10. Plott, C.R.: A notion of equilibrium under majority rule. Am. Econ. Rev. 57, 787–806 (1967)
11. Schofield, N.: The Spatial Model of Politics. Routledge, Abingdon (2007). 320p.
12. Saari, D.G.: Geometry of Voting. Springer, Heidelberg (2012). https://doi.org/10.1007/978-3-642-48644-9. 373p.

Intelligent Group Decision Making and Consensus Process

PredictRV: A Prediction Based Strategy for Negotiations with Dynamically Changing Reservation Value

Aditya Srinivas Gear$^{(\boxtimes)}$(iD), Kritika Prakash$^{(\boxtimes)}$(iD), Nonidh Singh$^{(\boxtimes)}$(iD),
and Praveen Paruchuri$^{(\boxtimes)}$(iD)

IIIT - Hyderabad, Hyderabad, India
{aditya.srinivas,kritika.prakash}@research.iiit.ac.in,
nonidh.singh@students.iiit.ac.in, praveen.p@iiit.ac.in

Abstract. Negotiation is an important component of the interaction process among humans. With increasing automation, autonomous agents are expected to take over a lot of this interaction process. Much of automated negotiation literature focuses on agents having a static and known reservation value. In situations involving dynamic environments e.g., an agent negotiating on behalf of a human regarding a meeting, agents can have a reservation value (RV) that is a function of time. This leads to a different set of challenges that may need additional reasoning about the concession behavior. In this paper, we build upon Negotiation algorithms such as ONAC (Optimal Non-Adaptive Concession) and Time-Dependent Techniques such as Boulware which work on settings where the reservation value of the agent is fixed and known. Although these algorithms can encode dynamic RV, their concession behavior and hence the properties they were expected to display would be different from when the RV is static, even though the underlying negotiation algorithm remains the same. We, therefore, propose to use one of Counter, Bayesian Learning with Regression Analysis or LSTM model on top of each algorithm to develop the PredictRV strategy and show that PredictRV indeed performs better on two different metrics tested on two different domains on a variety of parameter settings.

Keywords: Automated negotiation · Dynamic reservation value · Belief update

1 Introduction

Negotiation is an important component of interaction process among humans [18,19,22]. A lot of negotiation literature assumes that we have a good amount of information about our own choices [10,15] and reservation value (**RV**), while not knowing our opponents preferences [4,5,12]. Note that RV refers to the utility of a bid in the negotiation, below which we would not be willing to accept any bid. Reasons for not accepting a bid whose utility is below RV can be due to a

© Springer Nature Switzerland AG 2020
D. C. Morais et al. (Eds.): GDN 2020, LNBIP 388, pp. 135–148, 2020.
https://doi.org/10.1007/978-3-030-48641-9_10

better BATNA - Best Alternative to Negotiated Agreement [6] (so RV may be set to BATNA) or that the agent receives a utility that is not good enough for the agent to accept. In settings where the environment is dynamic, there can be situations where our RV can change with time (while the preference profile is static) [16]. We may not know how the changes would pan out e.g., an agent acting on behalf of a meeting attendee may have varying estimates on when the human may arrive for the meeting [3,23]. Dynamicity of RV can, therefore, throw additional challenges when we are unaware of the nature of changes (which is different from RV changing because of a discount factor where the change is computable). Bids that simply react to the dynamicity may not be sufficient since they can change in a random fashion and result in lower utility. For example, it can be hard to agree on a meeting time if an agent acting on behalf of a human declares that the human would arrive in 30 min and then re-declares in a short period that the human would arrive in 10 min and then quickly change to say 20 min even though the agent may simply be acting based on its belief of when the human would arrive.

1.1   Related Work

Making concessions to reach an agreement is an important part of the negotiation process [8,14,20]. There are a variety of ways in which negotiating agents can concede. One such category of techniques is Time-Dependent Tactics (TDT's) [7,9] e.g., Boulware and Conceder agents. [1] presents an Optimal Non-Adaptive Concession (ONAC) algorithm with incomplete information where time pressure (amount of time to deadline) is a primary criterion to influence the concession behavior. Negotiation algorithms such as ONAC and Boulware [1] work on settings where RV of the agent is fixed and known. Although these algorithms can work with (or be modeled as a function of) a dynamic RV, their concession behaviors can have a lot more randomness or fluctuations compared to when they have a static RV. For purposes of a more stable bidding behavior, the agent should, therefore, make choices based on predicted (RV) values. While the quality of agreement is a default metric used in negotiations, popular negotiation frameworks such as the Genius platform [17] do not support the modeling of dynamic RV. We, therefore, had to develop a simple negotiation simulator that can encode dynamic RV. In addition to the quality of agreement, we use Prediction as an additional metric to evaluate the concession behavior.

We propose to use the following models on top of negotiation algorithms, to handle the effects of a dynamic RV: (a) **Counter model** [24], (b) **Bayesian learning with Regression Analysis** [25,26] and (c) **LSTM model**. All three models are present in literature and we adapt them here to work suitably with the different negotiation algorithms. While the paper builds on top of ONAC and Boulware algorithms, the procedure, in general, would be suitable to apply to algorithms that are sensitive to the dynamicity of RV (which results in fluctuations in bidding). Given that the models help to predict the RV to reduce the effect of dynamicity, we refer to the new strategy as **PredictRV**.

Rest of the paper is organized as follows: Sect. 2 presents an overview of the negotiation model and two negotiation algorithms namely ONAC and Boulware with static RV. Section 3 presents a dynamic RV version of the negotiation model and the ONAC and Boulware algorithms. In addition, it introduces the PredictRV strategy and presents three methods used to make predictions over the dynamic RV namely Counter, Bayesian Learning with Regression Analysis and LSTM based prediction. Section 4 showcases the working of the three prediction methods via an example when faced with dynamic RV. In Sect. 5, we present a variety of experiments on two different domains to evaluate the performance of the PredictRV strategy. Section 6 presents the conclusions of the paper.

2   Static RV

2.1   Negotiation Model

The negotiation model we use follows the alternating offers protocol [21] for a bilateral negotiation: Consider two agents A and B with utility functions $U_A(z)$ and $U_B(z) \in [0,1]$ where z belongs to the set of all possible negotiation outcomes for a domain D. The RV's for the agents are rv_A and $rv_B \in [0,1]$. The agents will propose offers with utility higher than their own RVs.

2.2   Utility Generation for ONAC Algorithm

The ONAC algorithm [1] aims to construct optimal concession strategies against specific classes of acceptance strategies [2]. It applies sequential decision techniques to find analytical solutions that optimize the bidders expected utility, given certain strategy sets of the opponent. The ONAC solution was found to significantly outperform state of the art approaches in terms of obtained utility. As shown in [1], the utility of the ONAC bid is computed by taking into account the probability of acceptance of the bid (x, bid of agent A) by the opponents where the agents have opposing preferences.

$$U_j = U_{j+1} + \max_{U(x) \geq rv_A} (U(x) - U_{j+1})(1 - U(x))$$

where $U(x) \in [0,1]$ and U_{j+1} is the utility of the bid proposed by the agent at round $(j+1)$, U_j is the utility at round j and x is a valid bid with utility greater than RV. This is a recurrence formula that gives the utility of the bid at each round, where rv_A is the RV for agent A and N is the deadline:

$$U_N = rv_A, \quad U_j = (\frac{U_{j+1} + 1}{2})^2, \ j \in \{1, 2, 3, ..., N - 1\} \tag{1}$$

2.3   Utility Generation for Boulware Algorithm

The Boulware algorithm is a TDT [7,9], which concedes considerably more as the negotiation deadline approaches. TDTs consist of a family of functions that represent an infinite number of possible. The formula for tactics, one for each

$$U_j = rv_A + (1 - rv_A) * (\frac{min(N - j, N)}{N})^{\frac{1}{\beta}} \quad (2)$$

value of β this family of functions is as follows where j is the jth round and β should be in the range $(0, 1)$ for Boulware.

3   Dynamic RV: The PredictRV Strategy

3.1   Negotiation Model

The negotiation model remains the same as for the static RV case with the following difference: Since agent A's RV is dynamic, it is represented as $rv_A(t)$ (and $rv_B(t)$ for B for generality).

3.2   ONAC for Dynamic Reservation Values

To model dynamic RV we assume that the value of RV is drawn from an unknown probability distribution, and in each round, agent A receives a signal $rv_A(t)$ drawn from that distribution. PredictRV attempts to predict this probability distribution (p.d.) and incorporate it into a negotiation algorithm (ONAC here). We assume that there is no noise in the signal $rv_A(t)$, hence it corresponds to the actual RV at time step t. The PredictRV recurrence formula would be:

$$U_N = rv_A(t), \text{ where t} = j \text{ at round } j$$
$$U_j = (\frac{U_{j+1} + 1}{2})^2, \ j \in \{1, 2, 3, ...N - 1\} \quad (3)$$

For a dynamic RV, the value to bid will no longer be determined using Eq. (1). Instead, we first need to assign the new RV to U_N and then re-compute for U_j as shown in Eq. (3).

3.3   Boulware for Dynamic Reservation Values

The Boulware algorithm present in negotiation literature assumes a static RV. For Boulware that works with dynamic RV, utilities can be generated using the following function:

$$U_j = rv_A(j) + (1 - rv_A(j)) * (\frac{min(N - j, N)}{N})^{\frac{1}{\beta}}, \text{ at round } j \quad (4)$$

3.4   Illustrative Example

Consider a toy example, where the RV can be either 0.1 or 0.9 and it changes randomly every 2 rounds for a total of 100 rounds. Figure 1 shows the concession curves obtained by using the ONAC and Boulware algorithms. The x-axis of each figure shows the number of rounds from 0 to 100 while the y-axis shows the utility values ranging from 0 to 1. The utilities of the bids at each round are computed using Eq. (3). The figures show that the concession curves are not monotonic due to the dynamic nature of the RV, which results in to and fro concessions being made, where peaks correspond to RV of 0.9 and troughs correspond to 0.1.

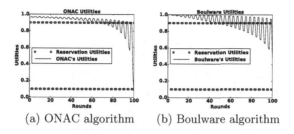

(a) ONAC algorithm   (b) Boulware algorithm

Fig. 1. Utilities obtained by using ONAC and Boulware algorithms

3.5   Steps of Strategy for PredictRV

Given a negotiation algorithm (like ONAC or Boulware):

1. Generate hypotheses about the RV and assign weights to each hypothesis. Compute the utility for each hypothesis T_{x_i} [by setting U_N as the utility of the hypothesis and plugging in Eq. (3)].
For each round j from 1 to N (no. of rounds):
2. Update weights of hypotheses based on the rv_A at that round i.e., $rv_A(j)$ [using Counter, Bayesian or LSTM approaches presented below]
3. Using the utility computed for each hypothesis in Step (1), we now compute the utility of the bid [Using one of Eqs. (6) or (13)].

End of for

To generate hypotheses (first step), we divide the range between which the RV can vary, into n number of intervals I_i for $i \in \{1, 2, 3, ...n\}$. A suitable point x_i is selected as a representative value for each interval I_i. If the RV falls within an interval, it is classified as having the utility of the point that represents the interval. We then compute negotiation algorithm utilities, $T_{x_i} = \langle U_1(x_i), U_2(x_i), ..., U_N(x_i) \rangle$ using Eq. (3). At the start, all hypotheses are equally likely, hence each hypothesis is initialized with a probability $\frac{1}{n}$ i.e., uniform distribution over hypotheses. As the negotiation progresses we may have a better prediction over the hypotheses based on the past RVs, hence the probability distribution would change. The second step of the PredictRV strategy is to update the weights of the hypotheses as the new round starts. How the weights are updated depends on the actual procedure we use namely Counter, Bayesian Learning or LSTM models presented below.

3.6   Counter Learning

In the Counter based learning procedure, the count for each hypothesis is initialized as $c_{x_i} = 0$, where $i \in \{1, 2, 3, ...n\}$. At a new round j, we obtain a new RV. As step 2 of PredictRV, using the new RV we update the counter for the hypothesis that corresponds to the new RV. We re-compute the probability for each interval as follows:

$$p_{x_i} = \frac{c_{x_i}}{\sum_{i=1}^{n} c_{x_i}}, \; i \in \{1, 2, 3, ...n\} \tag{5}$$

As step 3 of PredictRV, using the probabilities computed on different intervals we compute the utility U_j to be bid by PredictRV as:

$$U_j = \sum_{i=1}^{n} p_{x_i} * T_{x_i j}, \; j \in \{1, 2, 3, ...N\} \tag{6}$$

3.7   Bayesian Learning with Regression Analysis (BLRA)

In the BLRA procedure presented in [25], the learning agent i has a belief about the p.d. of its opponent's negotiation parameters (i.e., the deadline and RV). As shown in step 1 of PredictRV, we have a belief over the hypothesis of our own (dynamic) RV. By keeping track of the history of values obtained for RV so far and comparing it with fitted estimates derived from a regression analysis, the agent can revise its belief over the hypothesis by using a Bayesian updating rule and can correspondingly adapt its concession strategy.

Regression Analysis. As the negotiation proceeds [25], utility u_t for a TDT decreases according to the following decision function:

$$u_t = 1 - \left(\frac{t}{T}\right)^{\beta} \tag{7}$$

where T is the deadline and β is the concession parameter. We adopt this terminology to express in terms of agent A's own dynamic RV. We assume RV to be 0 at the start of the negotiation and vary according to Eq. (7).

$$u_t = u_0 + (u_T - u_0)\left(\frac{t}{T}\right)^{\beta} \tag{8}$$

where u_T is the RV at the deadline and u_0 is the RV at the start. For every round, we receive an RV for that round. We compute the regression line (fitted utilities) $\hat{RV}_{t_b} = \{\hat{u}_0, \hat{u}_1, \hat{u}_2, ..., \hat{u}_{t_b}\}$ based on the historical RVs, $RV_{t_b} = \{u_0, u_1, u_2, ..., u_{t_b}\}$ until round t_b as follows:

Step 1: Generate the hypotheses and initialize its probabilities as mentioned in Sect. 3.5 (Steps of strategy) with x_i representing the utility of each hypothesis.
Step 2: Based on Eq. (8), we use the following power regression function to calculate the regression curve:

$$\hat{u}_t = u_0 + (x_i - u_0)\left(\frac{t}{N}\right)^{\beta} \tag{9}$$

where N is the deadline. Next, β is calculated using Eq. (10) (as proposed in [25]):

$$\beta = \frac{\sum_{k=1}^{t_b} t_k^* u_k^*}{\sum_{k=1}^{t_b} t_k^{*2}}, \text{ where } u_k^* = ln\left(\frac{u_0 - u_k}{u_0 - x_i}\right), t^* = ln\left(\frac{t}{N}\right) \tag{10}$$

Step 3: Based on the calculated regression curve given by Eqs. (9) and (10), the fitted RVs, $\hat{RV}_{t_b}$ would be $= \{\hat{u}_0, \hat{u}_1, \hat{u}_2, ..., \hat{u}_{t_b}\}$ at each round (where $\hat{u}_0 = u_0$).
Step 4: We now calculate the non-linear correlation between RV_{t_b} and the fitted RVs $\hat{RV}_{t_b}$. The coefficient of non-linear correlation γ is given by Eq. (11), where $\overline{u}$ and $\overline{\hat{u}}$ are the average of all the historical and fitted RVs respectively:

$$\gamma = \frac{\sum_{k=1}^{t_b}(u_k - \overline{u})(\hat{u}_k - \overline{\hat{u}})}{\sqrt{\sum_{k=1}^{t_b}(\hat{u}_k - \overline{\hat{u}})^2 \sum_{k=1}^{t_b}(\hat{u}_k - \overline{\hat{u}})^2}}, \gamma_{new} = \frac{\gamma + 1}{2} \tag{11}$$

Step 5: Parameter γ $(-1 \leq \gamma \leq 1)$ is used for evaluating resemblance between chosen (x_i) and real RVs (u_t). To use γ as a probability to perform belief update in Bayesian Learning, we normalize it to [0,1] (γ_{new} in Eq. (11)).

Bayesian Learning

Step 1: Bayesian Learning can be used if we have a hypothesis about the prediction. Belief about p.d. of these hypotheses can be revised through a posterior probability by observing the RV. Each hypothesis H_i represents that it would be the possible RV at the end of negotiation. The prior p.d., denoted by $P(H_i)$, $i \in (1, 2, 3, ..., n)$ signifies the agent's belief about the hypothesis i.e., how likely the hypothesis matches the RV at the end of the negotiation.
Step 2: The agent can initialize the p.d. over hypotheses based on some prior information if available, otherwise a uniform distribution $P(H_i) = \frac{1}{n}$ is assigned. During each round of negotiation t_b the probability of each hypothesis would be computed using the Bayesian updating rule in Eq. (12):

$$P(H_i|RV) = \frac{P(H_i)P(RV|H_i)}{\sum_{k=1}^{n} P(RV|H_k)P(H_k)} \tag{12}$$

Step 3: The observed outcome here is historical RVs $RV_{t_b} = \{u_0, u_1, u_2, ..., u_{t_b}\}$. As presented in [25], the agent will update the prior probability $P(H_i)$ using the posterior probability $P(H_i|RV_{t_b})$, thus a more precise estimate is achieved using Eq. (12).
Step 4: As presented in [25], conditional probability $P(RV_{t_b}|H_i)$ is obtained by comparing the fitted points $\hat{RV}_{t_b}$ on the regression line based on each selected RV x_i, with the historical RVs RV_{t_b}. The more correlated fitted RVs are with historical RVs, the higher $P(RV_{t_b}|H_i)$ will be.
Step 5: Difference between the regression curve and the real RV sequence can be indicated by the non-linear correlation coefficient γ_{new}. Thus, we can use the value of γ_{new} as the conditional probability $P(RV|H_i)$ in Eq. (12). The learning approach will increase the probability of a hypothesis when the RV selected (x_i) is most correlated with the RV at the end of the negotiation. As mentioned in step 4 of PredictRV, using the probabilities on different intervals, we compute the utility at that round as:

$$U_j = \sum_{i=1}^{n} P(H_i) * T_{x_ij}, \quad j \in \{1, 2, 3, ...N\} \tag{13}$$

3.8   LSTM Based Prediction

LSTM (Long-Short Term Memory) [13] is a popular recurrent neural network architecture to perform deep learning tasks and is useful in time-series prediction. The negotiation problem introduced here can be modeled as a time series prediction task wherein the agent learns more information as the negotiation progresses. We, therefore, propose to use an LSTM based approach to predict the RV at the last time step n of the

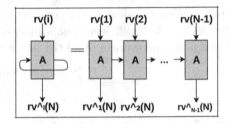

Fig. 2. LSTM architecture

negotiation, using time-series forecasting. As shown in Fig. 2, the input at each time step t for LSTM is $RV(t)$ (i.e., RV provided by the environment at t). Note that there exists a single LSTM cell A to which input is fed repeatedly (one value at every time step) along with the output of the previous time step. Output at t is the predicted value for RV at the last time step n denoted by $\hat{RV}_t(n)$. The LSTM is trained using a mean squared error loss function and learns to predict better as the number of epochs increases. There are n hypotheses in our problem whose probability is updated every time step based on the predicted RV for the last time step $\hat{RV}$. This is similar to Counter model where we identify the interval the $\hat{RV}$ falls into and increase the count of that hypothesis by 1 (Eq. (5)). Using the probabilities for different hypotheses we compute the utility to be bid by PredictRV (Eq. (6)).

4   Example Continued

The rest of the example is explained using the ONAC-D algorithm (ONAC-D is ONAC strategy without any changes applied to Dynamic RV). Figure 4 shows the utility values generated by Counter, BLRA and LSTM models computed using Eqs. (6) and (13) respectively. The x-axes shows the number of rounds from 0 to 100 while the y-axes shows the utility values ranging from 0 to 1.

Figure 3 shows the belief plots for the three models. A belief plot shows how the belief in a particular hypothesis changes as the rounds progress. The figure shows two plots corresponding to the two hypotheses that the RV is 0.1 (hypothesis 0.1) and 0.9 (i.e., hypothesis 0.9). The x-axes for both the figures show the number of rounds from 0 to 100 and the y-axes show the probability of belief in the hypothesis that the figure represents e.g., a y-axis value of 0.3 in figure on left implies that an algorithm believes that the RV is 0.1 with a probability 0.3 which implies that other hypotheses are true with rest of the probability (in this case only other hypothesis is hypothesis 0.9). The belief plots show that:

(a) For hypothesis 0.1, while Counter stays close to middle (probability of 0.5), BLRA and LSTM are more clear in their belief for this hypothesis (former

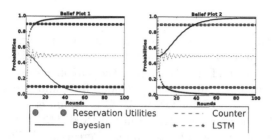

Fig. 3. Belief plots for two hypotheses

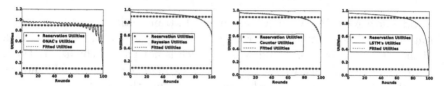

Fig. 4. Algorithms with their fitted curves

converges to close to 0 while the latter converges to close to 1 probability and stay with these probabilities once converged) showing the inherent differences between the models. (b) Counter converges quickly to a belief of 0.5 since RV alternates between the hypotheses every 2 steps, hence the count is more or less balanced. (c) For BLRA, belief in hypothesis 0.1 converges close to 0 since it is not just the count but the time when the RV changes come into play here. (e) For LSTM, belief in hypothesis 0.1 converges to close to 1 faster than other models, however to the opposite belief of BLRA for this example. The outcome utility for $\langle$ONAC-D, Counter$\rangle$ is $\langle 0.5, 0.5 \rangle$, $\langle$ONAC-D, Bayesian$\rangle$ is $\langle 0.25, 0.75 \rangle$ and $\langle$ONAC-D, LSTM$\rangle$ is $\langle 0.6, 0.4 \rangle$.

5   Experiments

5.1   Setup for the Experiments

We have a number of hypotheses, number of rounds of negotiation N and update rate (frequency of change in RV) as the parameters of our algorithm. N is fixed to 100 for all experiments. Experiments were performed on the Fire Disaster Response and Meeting Scheduling domains. In both these domains, the agent is faced with a dynamic RV. For purposes of experimentation, we model the dynamic RV using a Markov chain model [we omit the specifics of our modeling due to space constraints]. The number of hypotheses vary across the domains. Update rate of RV is varied among the values $\{2, 5, 10, 20, 50\}$. We run each experiment for 100 iterations keeping the parameters constant. X-axis shows the (hypothesis, update rate) while y-axis shows the respective metric in each plot.

5.2   Metrics

1) **Outcome Utility Metric:** We run negotiations for agent A vs agent B, where A uses one of ONAC-D or Boulware-D and B is PredictRV strategy (Counter, BLRA or LSTM). We average the outcome over 100 iterations and compute the outcome utility for each UpdateRate and hypothesis (averaged utility represented as OD for ONAC-D, C for Counter, B for BLRA and L for LSTM). We then compute the utility of PredictRV w.r.t ONAC-D using Eq. (14) (represented in graphs as Average Percentage Utility):

$$\text{percentage utility of i} = \frac{i - OD}{OD} * 100, i \in \{C, B, L\} \qquad (14)$$

2) **Prediction Metric:** We allow each model to train until the end of the negotiation (N rounds). At the last round N, we have an RV predicted by each of the models i.e $\hat{RV}$ for round $N + 1$ (which is not part of the negotiation). For each of the models, we then compute the difference between $\hat{RV}$ and the actual RV at round $N + 1$ which is used to capture the quality of prediction. This value is averaged over 100 iterations where a lower difference in average value implies a better prediction.

5.3   Fire Disaster Response

Consider a forest fire where the fire can spread quickly in any of the 4 directions i.e., North, South, East or West. Assume that the forest is modeled as a grid of size n_1*n_1 [11]. Fire fighting units (local units) are dispatched to many locations to fight the fire. The commander in charge has a global picture of the fire and wants to reduce the resources given to each local unit. The local unit leader (modeled as agent) would like to negotiate with the commander to obtain higher (than minimal needed to just put off the fire) number of resources to stop the fire quickly at the local point. Given that the direction of fire changes in different time steps, the RV is dynamic i.e. changes with time.

We operationalize the experimental parameters as follows: A negotiation is being carried out with N (=100) as the deadline. The parameters here are number of hypotheses, the update rate of the RV and grid size. The number of hypotheses are varied among the values {2, 4} i.e., {North, South} or {North, West, East, South} directions with {0.75, 0.15} and {0.75, 0.57, 0.32, 0.15} (corresponding to number of resources {12, 10, 7, 4}) as the values for RV. Experiments were performed with the local location start point, as a random point around the center of the grid (up to a radius of 4 units from the center).

Figure 5 shows two plots corresponding to ONAC and Boulware with a random start point for fire with grid size 100. Both plots show that the values for outcome utility metric (Sect. 5.2) for PredictRV are higher than for ONAC-D or Boulware-D respectively e.g., plot (b) of Fig. 5 shows that the Average Percentage Utility for BLRA varies from 50% at the lowest to 95% at the highest. The plot also shows that the overall Average Percentage Utility across all the intervals and update rates for BLRA is 71% while it is 61% for Counter and −4.4% for LSTM.

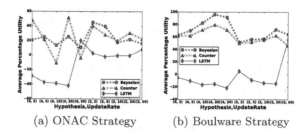

(a) ONAC Strategy   (b) Boulware Strategy

Fig. 5. Outcome utility in fire domain with random start points

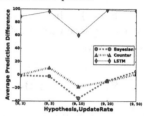

To showcase the statistical significance of the outcome utility results presented in Fig. 5 of the paper, we performed a paired valued t-tests for the following settings (where O: ONAC, Bo: Boulware, Ba: Bayesian, C: Counter, L: LSTM): In PredictRV experiments for O: Ba vs. O, C vs. O, L vs. O, Ba vs. C, Ba vs. L, C vs. L. In PredictRV experiments for Bo: Ba vs. Bo, Co vs. Bo, L vs. Bo, Ba vs. C, Ba vs. L, C vs. L. If the calculated P-value is less than 0.05, it means that statistically the mean difference (in outcome utility as shown in Fig. 5) between the

Fig. 6. Prediction in the meeting domain

paired observations is different. Our testing showed that mean values of outcome utility for Bayesian vs. Counter does not have significant difference statistically (both for ONAC and Boulware i.e., 2 tests). All the other (10) tests, showed that the differences in (the averaged) outcome utility are statistically different.

5.4   Meeting Scheduling Domain

For brevity purposes, we present the gist of the domain here: We operational-ized parameters for this domain from the E-Elves [23] application. The param-eters for the algorithm are the update rate of the RV, delay intervals and number of hypotheses. There are 9 possible delay intervals we consider here i.e $\{5, 10, 15, 20, 25, 30, 35, 40, 45\}$ min. A delay interval of 10 min means that a meeting supposed to start at 10 am is now rescheduled to start at 10:10 am. For this domain each hypothesis corresponds to a delay interval, hence 9 hypotheses correspond to 9 delay intervals. The overall value of the meet is computed as below:

$$Delaycost = (delay^{alpha}) * 2, Value\,of\,the\,meet = 200$$
$$Overall\,value = Value\,of\,meet - Delaycost \tag{15}$$

where $delay$ is delay w.r.t. the scheduled starting time and $alpha \in \{1.0, 1.2, 1.4, 1.6\}$. Utility of the hypothesis is calculated by normalizing the reward obtained using Eq. (15). The prediction measurement for the meeting domain is shown in Fig. 6. Experiments for measuring prediction for the meet-ing domain were performed with the following summary (we skip graphs due to

Metrics	Meeting Domain			Fire Random Domain		
	Counter	Bayesian	LSTM	Counter	Bayesian	LSTM
ONAC Outcome Utility	1(20%)	2		2	1(5%)	
Boulware Outcome Utility	2	1(7.2%)		2	1(13.04%)	
Prediction	2		1 (89.17%)	2		1 (33.91%)

Fig. 7. Relative performance table (ranks)

space issues): The overall average percentage prediction across all the intervals and the update rates for Counter, BLRA and LSTM are -3.05, -8.27 and 88.12 respectively.

5.5 Summary of the Experiments

In Fig. 7, 1 signifies the best performing model and 2 signifies the second-best performing model for a given metric and domain, $x\%$: how much better the best model is relative to the second-best model. Formulation: $a =$ metric value of best model, $b =$ metric value of 2nd best model. For the outcome utility relative performance $= 100 * \frac{(a-b)}{a}$, $(a > b)$. For the prediction metric, relative performance $= 100 * \frac{(b-a)}{b}$, $(b > a)$.

Explanation: For each of the metrics, we measure the relative value of the best performing model w.r.t the second-best performing model for each domain. For example, in the Fire Random domain for the ONAC algorithm, BLRA is 5% better than Counter on outcome utility metric and LSTM is 33.91% better than Counter on the prediction metric.

6 Conclusions

We introduced the PredictRV strategy which uses one of Counter, BLRA or LSTM learning models that predict over the dynamic RV to perform a better negotiation. Our results show that: a) For Outcome Utility: the BLRA model performs slightly better than Counter although the difference is not statistically significant. b) For Prediction metric: LSTM is the best performing model while Counter performs next best. c) Outcome utility is the standard metric that is used to evaluate negotiations. Given that both BLRA and Counter methods perform well on this metric, they can be tested for the specific use case needed and one of them picked based on the insights obtained. In summary, the key novelty of our work is that we enhance the ability of current negotiation algorithms to handle dynamic RV. The problem can be more general where only an indicator function for the RV is available rather than the actual value at each update step as assumed here. Popular negotiation platform such as Genius allows us to encode static RV currently – we believe this work takes a significant step towards dealing with challenges in handling dynamic RV.

References

1. Baarslag, T., Hadfi, R., Hindriks, K., Ito, T., Jonker, C.: Optimal non-adaptive concession strategies with incomplete information. In: Fukuta, N., Ito, T., Zhang, M., Fujita, K., Robu, V. (eds.) Recent Advances in Agent-based Complex Automated Negotiation. SCI, vol. 638, pp. 39–54. Springer, Cham (2016). https://doi.org/10.1007/978-3-319-30307-9_3
2. Baarslag, T., Hindriks, K., Jonker, C.: Effective acceptance conditions in real-time automated negotiation. Decis. Support Syst. **60**, 68–77 (2014)
3. Chalupsky, H., et al.: Electric elves: agent technology for supporting human organizations. AI Mag. **23**(2), 11 (2002)
4. Choi, S.P., Liu, J., Chan, S.P.: A genetic agent-based negotiation system. Comput. Netw. **37**(2), 195–204 (2001)
5. Coehoorn, R.M., Jennings, N.R.: Learning on opponent's preferences to make effective multi-issue negotiation trade-offs. In: Proceedings of the 6th International Conference On Electronic Commerce, pp. 59–68. ACM (2004)
6. Crump, L., Moon, D.: Precedents in negotiated decisions: Korea-Australia free trade agreement negotiations. Negot. J. **33**(2), 101–127 (2017)
7. Faratin, P., Sierra, C., Jennings, N.R.: Negotiation decision functions for autonomous agents. Robot. Auton. Syst. **24**(3–4), 159–182 (1998)
8. Fatima, S., Kraus, S., Wooldridge, M.: Principles of Automated Negotiation. Cambridge University Press, Cambridge (2014)
9. Fatima, S., Wooldridge, M., Jennings, N.R.: Optimal negotiation of multiple issues in incomplete information settings. In: Proceedings of the Third International Joint Conference on Autonomous Agents and Multiagent Systems, vol. 3, pp. 1080–1087. IEEE Computer Society (2004)
10. Fatima, S.S., Wooldridge, M., Jennings, N.R.: An agenda-based framework for multi-issue negotiation. Artif. Intell. **152**(1), 1–45 (2004)
11. Heyman, D.P., Sobel, M.J.: Stochastic Models in Operations Research: Stochastic Optimization, vol. 2. Dover Publication, New York (1982). Courier Corporation
12. Hindriks, K., Tykhonov, D.: Opponent modelling in automated multi-issue negotiation using bayesian learning. In: Proceedings of the 7th International Joint Conference on Autonomous Agents and Multiagent Systems, vol. 1, pp. 331–338. International Foundation for Autonomous Agents and Multiagent Systems (2008)
13. Hochreiter, S., Schmidhuber, J.: Long short-term memory. Neural Comput. **9**(8), 1735–1780 (1997)
14. Jennings, N.R., Faratin, P., Lomuscio, A.R., Parsons, S., Wooldridge, M.J., Sierra, C.: Automated negotiation: prospects, methods and challenges. Group Decis. Negot. **10**(2), 199–215 (2001)
15. Kraus, S.: Strategic Negotiation in Multiagent Environments. MIT Press, Cambridge (2001)
16. Li, M., Vo, Q.B., Kowalczyk, R., Ossowski, S., Kersten, G.: Automated negotiation in open and distributed environments. Expert Syst. Appl. **40**(15), 6195–6212 (2013)
17. Lin, R., Kraus, S., Baarslag, T., Tykhonov, D., Hindriks, K., Jonker, C.M.: Genius: an integrated environment for supporting the design of generic automated negotiators. Comput. Intell. **30**(1), 48–70 (2014)
18. Raiffa, H.: The Art and Science of Negotiation. Harvard University Press, Cambridge (1982)
19. Rosenfeld, A., Kraus, S.: Predicting human decision-making: from prediction to action. Synth. Lect. Artif. Intell. Mach. Learn. **12**(1), 1–150 (2018)

20. Rosenschein, J.S., Zlotkin, G.: Rules of Encounter: Designing Conventions for Automated Negotiation Among Computers. MIT Press, Cambridge (1994)
21. Rubinstein, A.: Perfect equilibrium in a bargaining model. Econom.: J. Econom. Soc. **50**, 97–109 (1982)
22. Sandholm, T., Lesser, V.R., et al.: Issues in automated negotiation and electronic commerce: extending the contract net framework. ICMAS **95**, 12–14 (1995)
23. Scerri, P., Pynadath, D.V., Tambe, M.: Towards adjustable autonomy for the real world. J. Artif. Intell. Res. **17**(1), 171–228 (2002)
24. Scheaffer, R.L., Young, L.: Introduction to Probability and its Applications. Cengage Learning, Boston (2009)
25. Yu, C., Ren, F., Zhang, M.: An adaptive bilateral negotiation model based on bayesian learning (2013)
26. Zeng, D., Sycara, K.: Bayesian learning in negotiation. Int. J. Hum.-Comput. Stud. **48**(1), 125–141 (1998)

Inferring Personality Types for Better Automated Negotiation

Sai Naveen Pucha[(✉)] and Praveen Paruchuri[(✉)]

International Institute of Information Technology, Hyderabad, India
sainaveen.pucha@research.iiit.ac.in, praveen.p@iiit.ac.in

Abstract. Automated negotiation between computational agents or between agents and humans has been a subject of active research with a focus on obtaining better quality solutions within reasonable time frames. The critical issue negotiators face during automated negotiation is that a negotiator may not always know the personality type of the opponent. Studies show that having information about the opponent improves the outcome of negotiation in general. However, unless there is prior knowledge, learning the opponent type in the limited amount of time or number of rounds in a negotiation is a difficult task. In this paper, we use a Partially Observable Markov Decision Process (POMDP) based modeling to perform better modeling of the opponent personality type. In particular, we focus on modeling the opponent into four different types to showcase that a better understanding of personality type can improve the outcome of automated negotiation. Our experiments performed using data sets generated from the IAGO software showcase that we indeed obtain better negotiation outcomes with a higher classification accuracy of the opponent personality type.

Keywords: Automated negotiation · Personality type · Belief tracking · POMDP

1 Introduction

The topic of automated negotiation has received wide attention in the literature due to its importance as one way of communication between agents and also between agents and humans [3,17,18]. In a general automated negotiation, an agent may need to negotiate with other agents or humans whose (personality) type may not be known [2,24]. However, knowing the types of other players can provide significant advantages to the agent in many situations in terms of having a better idea of how to approach or tailor the negotiation [6,9,10], potentially leading to better outcomes. The offer-counteroffer paradigm [8,12] is a popular way of modeling automated negotiations. In this study, we deviate from the offer-counteroffer paradigm since the other negotiators can take actions for information exchange and not just to make offers and counteroffers.

In the business world, people bring different negotiation styles and strategies to the bargaining table, based on their different personalities, experiences, and

© Springer Nature Switzerland AG 2020
D. C. Morais et al. (Eds.): GDN 2020, LNBIP 388, pp. 149–162, 2020.
https://doi.org/10.1007/978-3-030-48641-9_11

beliefs about the negotiation. When people with different negotiation styles are involved in a negotiation, the results can be a lot more unpredictable. On knowing more about the other parties, negotiation styles can have a significant impact on the way we plan our negotiation. Individual differences in social motives and preference for certain outcomes when agents interact can strongly affect how they would approach the negotiation [5]. Drawing on social motives that drive human behavior, [25] and others identify four basic personality types:

- **Individualists** concentrate primarily on maximizing their outcomes and do not show much concern for opponent outcomes.
- **Cooperators** focus on maximizing both their own and other's outcome. Cooperators tend to opt more for value creation strategies such as exchanging information than individualists.
- **Competitives** are motivated to maximize the difference between their own and other's outcome. Because of their strong desire to "win big", competitive agents tend to engage in behavior that is self-serving.
- **Altruists** are a rare breed of negotiators that strive to maximize other's outcome rather than their own. Although very few people may be pure altruists, most human negotiators behave altruistically under specific conditions such as when dealing with loved ones or with those who are less fortunate.

Given the above classification for personality type of negotiators, our goal is to build an automated negotiation agent that can adjust its belief regarding the type of opponents as the negotiation progresses and chooses appropriate actions to reach a high-quality negotiation agreement. There is quite some work on modeling and prediction of opponent type in negotiation literature [1,2] including usage of formal models such as game theory [4,7] and Bayesian learning [26,27] among others. However, there is relatively less work on analyzing the personality type of an opponent within the computation model of negotiation. POMDPs [11] have been used earlier in literature for modeling social interaction among humans in situations where information of the opponent has to be known [21]. More recently, preference elicitation in negotiation using the Gaussian uncertainty model has been developed in order to optimize the negotiation outcomes [14,15]. [22] presents a POMDP based model for the development of a strategic agent for human persuasion via the usage of argumentative dialogs.

The assumption we make in our work is that initially, the agent has no (or little) idea of the type of the other parties (hence we assume a uniform distribution over the possible types). Throughout the negotiation, the agent receives feedback and refines its belief accordingly. Given the characteristics, we would like to capture, prior work has shown that POMDP based modeling can be a good fit [11]. We focus on negotiation between two agents in the rest of the paper and present the POMDP framework for our negotiation problem as the next step, along with details of how it encodes our problem.

The rest of the paper is organized as follows: Sect. 2 presents details on how we model the Automated Negotiation problem as a POMDP. Section 3 describes usage of IAGO software to obtain a dataset that post some processing is used as input for the POMDP model. Section 4 performs a basic evaluation of our

POMDP model. In particular, the section compares the POMDP agent against corresponding MDP models with differing assumptions, performs reward function evaluation, and also studies the effect of misclassification of opponent type. Section 5 presents the experimental setups used and the results obtained. In particular, the section shows results on learning of opponent personality type in terms of classification accuracy, obtained via the POMDP belief updates. Section 6 presents the conclusions of our work.

2   Partially Observable Markov Decision Process Framework

Formally, POMDP can be defined by the tuple $\{S, A, T, \Omega, O, R\}$, where S is a finite set of states; A is a finite set of actions; $T(s, a, s')$ captures the probability of transitioning from state s to s' when taking action a; Ω is a finite set of observations; $O(s', a, o)$ is the probability of observing o when taking action a leads the agent to state s' and $R(s, a)$ represents the reward function, i.e., the reward obtained by taking action a, at state s. As part of modeling the POMDP, we also need to specify an initial belief where a belief state b is defined as a probability distribution over the set of states S. Once the negotiation problem is cast into the POMDP framework, many algorithms both heuristic and exact exist in the literature to find an approximate or optimal POMDP policy [11,20]. Note that a policy here refers to a mapping from a belief state b to an action a, for all the possible valid belief states.

2.1   Encoding a Negotiation Problem in the POMDP Framework

Using the ideas presented in prior work [18], we encode the negotiation problem into a POMDP tuple $\{S, A, T, \Omega, O, R\}$. As has been noted in prior works [18,19], the advantages of a POMDP-based modeling approach for negotiation are as follows: (a) POMDPs provide a natural way to capture the sequential nature of the negotiation process while reasoning about the new data observed (such as the actions of the other agent). (b) POMDPs enable to model and refine an agent's belief about other agents (in this work, the belief is over the personality type of the other agent).

A key issue we face here is that the negotiation transcripts typically have utterances, while a POMDP needs specific actions to be defined. To make the mapping of negotiation transcripts to a POMDP feasible, we first assign a set of codes to the actions of the players expressed in terms of dialogues [18]. Each utterance can correspond to one or more codes, and these codes form the action set that can be taken by each player. The POMDP would, therefore, be modeled in terms of codes. We now provide details of the POMDP encoding for a two-player negotiation.

2.2 State Space Definition

The state space of our POMDP has the following factors: <Type, MyProposal, OpponentProposal, MyAction, Observation, PreviousObservation>.

- **Type:** Modeled four personality types.
- **MyProposal:** The bid which the agent has to offer.
- **OpponentProposal:** The bid which the other player has to offer.
- **MyAction:** The action the agent has taken.
- **Observation:** The action which the other player takes, received by the agent as an observation.
- **PreviousObservation:** Observation of the agent in the last time step.

2.3 Action and Observation Set

There are two action categories from which the agents can choose. One category of actions is Information actions, while the other is Proposal actions. Information actions are those which do not involve a change in proposal values. Proposal actions, on the other hand, are used for changing the proposal value each agent makes. The negotiators can interact with each other using dialogues. The dialogue set that we use for experimentation is limited and is already a part of the IAGO software. The players get to pick their responses from this set.

We used the standard K-means clustering with tf-idf (term frequency–inverse document frequency) score as weights [23] to categorize each of the dialogues into 5 (code) categories. The tf-idf is a measure of the importance of a particular word with respect to the other words in a document. We assign the following five (action) codes to each of the five clusters generated by the K-means algorithm:

- **a1 - information** - This is a simple information exchange action.
- **a2 - same** - This indicates that the agent has the same value for the proposal again.
- **a3 - concede** - This means that the agents has conceded its proposal value.
- **a4 - agree** - This action indicates that the agent has agreed to the value/bid proposed by its opponent.
- **a5 - finish** - This action indicates that the negotiation has been completed/terminated.
- **a6 - NOOP** - This action has no impact on the negotiation process.

The key advantage of using action codes is that it makes the interaction amenable to being modeled using a POMDP due to their limited number. For this experiment, the rate at which the agent and the players concede is the same. The reservation value for the agent is dependent on its personality type and sets lower bound for acceptability of bid. Observation set (Ω) has the six codes that were part of the action set since observation for the agent is an action of the other player. In addition to the set of six codes, IAGO software also provides the emotion of the other player, which we model as part of the observations. In particular, we use four different emotions, namely Happy, Sad, Surprised, and

Angry, hence forming 20 action-emotion pairs as the observation set plus NOOP which leads to the following 21 observations in total: $a1 - happy$, $a1 - sad$, $a1 - angry$, $a1 - surprised$, $a2 - happy$, $a2 - sad$, $a2 - angry$, $a2 - surprised$, $a3 - happy$, $a3 - sad$, $a3 - angry$, $a3 - surprised$, $a4 - happy$, $a4 - sad$, $a4 - angry$, $a4 - surprised$, $a5 - happy$, $a5 - sad$, $a5 - angry$, $a5 - surprised$, $a6$

2.4   State Transition Function

The transition function represents the probability with which the agent reaches state s' when it takes action a from state s. The agent can take different actions from the action set, which can either be a proposal action or an information action. Upon taking these actions, the agent will make transitions to different states. The transitions in this domain are stochastic since information related to other players is part of our state, which we would not know beforehand.

2.5   Observation Function

The state in our POMDP model is a tuple, as described in subsect. 2.2. We follow the sequence of steps presented in [18] that show that when the state captures the previous observation, the observation function becomes deterministic. For explanation purposes, if ts is the current time step, we mathematically represent the observation function as follows [where $aa(ts)$ is the agent action taken and $oa(ts)$ is the opponent action taken at ts, i.e., the observation at ts]:

$$pr(obs(ts)|s(ts + 1), aa(ts)) = pr(obs(ts)|s(ts + 1))$$
$$= pr(obs| < Type, MyProposal(ts), OpponentProposal(ts),$$
$$MyAction(ts), Observation(oa(ts + 1)), PreviousObservation(oa(ts)) >)$$
$$= pr(obs(ts)|oa(ts))$$
$$= 1, \ if \ obs(ts) = oa(ts)$$
$$= 0, \ otherwise$$

2.6   Reward Function

The reward function is based on the personality type of the player. We introduced four personality types earlier, and each personality type has its reward function. As mentioned in subsect. 2.3, while emotions are part of the observations, in the current work, we do not condition reward values on the (observed) emotions. More specifically, the reward function for each personality type depends on whether a concession is made or not. For the Individualist, Cooperator and Competitor personality type, making a concession is not preferable for the agent; hence it is given a lower reward value. Please note that the meaning/value of concession by the same amount is different across the (3) personality types; hence the reward value is modeled accordingly. For example, between the Individualist and Cooperator agents, Individualists may never want to concede, but Cooperator will tend to concede at some time steps. For an Altruist, a concession is viewed as a positive since it helps its opponent, hence receives a high reward for a concede action.

2.7   Discount Factor

The value of the discount factor can vary from 0 to 1 in general. A discount factor close to 1 indicates that rewards in the distant future are of high priority while a value closer to 0 indicates that only immediate rewards are considered. We use a discount factor of 0.95 for our purposes.

3   Input Generation for the POMDP Model

We used the IAGO software [16] to obtain the input dataset. We then perform a processing step over this dataset to obtain state vectors in the form defined in Sect. 2.2. Please note that the dataset is not labeled on the personality types we model in this paper. We therefore went through the data set and labeled the data to include the personality types, e.g., if a player agent repeatedly agrees to a proposal made by the other agent without much bargaining involved, it is labeled as an Altruist. The state vectors of the POMDP are then built directly from the negotiation transcripts generated from the IAGO software.

4   Evaluation of the POMDP Agent

4.1   Sanity Check Experiment

In order to check whether the POMDP model we built is reasonable, we designed 2 MDP agents, namely MDP1 and MDP2. MDP1 doesn't capture the opponent personality type, whereas MDP2 knows the type of the opponent.

Figure 1 shows the results of the experiment, where the POMDP agent is evaluated using the MDP2 actions. There is a mapping step involved in this evaluation, since there is no observation feature in the MDP2 states. We perform this mapping using the remaining features of the state vector. Hence, it translates to a many to one mapping where multiple POMDP states can map to a single MDP2 state in which case the same optimal (MDP2) action is used for these states. The figure shows the number of timesteps on the x-axis and the Estimated Expected Total Rewards EETR [13] on the y-axis. From the figure, we deduce the following:

- EETR values for the POMDP and MDP2 agents are similar when the POMDP is provided information about the opponent's actions. Figure 1 shows the EETR values for POMDP and MDP2, however the lines overlap. We therefore represent MDP2 using a dotted green line and POMDP by an orange line with the + symbol for purposes of contrast.
- MDP1 obtains lesser EETR values, since it cannot make use of additional information about the opponent.

EETR is calculated as follows using Eq. 1:

$$V_t^*(s) = \max_a \left[R(s,a) + \gamma \sum_{s \in S} T(s,a,s') V_{t-1}^*(s') \right] \tag{1}$$

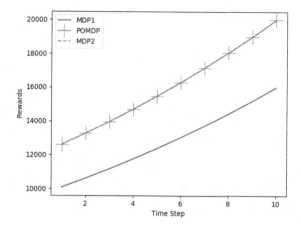

Fig. 1. Estimated expected total reward values for the computed policies of POMDP and the two MDP agents when given the same action sequences.

4.2 Reward Function Evaluation

The reward function is a component that is not present explicitly in the dataset, nor the data presents an obvious way to construct. As described in Sect. 2.6, the reward function is built, keeping in view the opponent types that need to be captured. With a well-designed reward function, the agent is expected to provide better EETR values if the reward function intended for that personality type is used. For example, if the POMDP uses the reward function of Individualist for an opponent who is Altruist due to incorrect modeling, it should be expected to return lesser EETR values than when it uses the correct reward function for Altruist. Figure 2 shows four plots, one per each personality type of the opponent. Each plot shows results for the experiment where we use the four different reward functions for a fixed personality type of the opponent [i.e., three incorrect and one correct reward function used]. Note that for purposes of this experiment, the state space of POMDP has only one opponent type information captured (instead of four as described above in State Space Definition), e.g., value for Type is Altruist in POMDP for plot (a), Cooperator in (b) and so on. The x-axis for each plot shows Time Steps, while the y-axis shows the reward obtained. From the plots, we can infer that when the correct reward function is used, the agent obtains better reward values as compared to when using incorrect one(s).

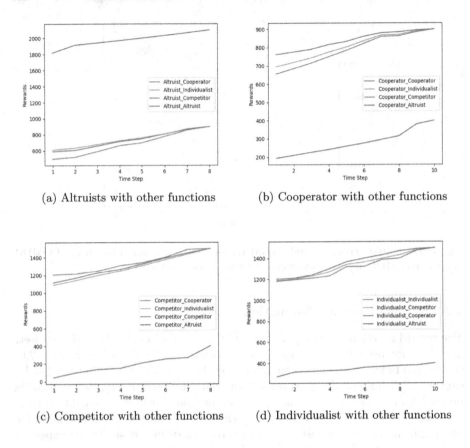

(a) Altruists with other functions   (b) Cooperator with other functions

(c) Competitor with other functions   (d) Individualist with other functions

Fig. 2. Each personality type tried with all the reward functions

4.3 Effect of Opponent Type Misclassification in Input Transcripts

This experiment while having similarities also differs from the previous experiment (on reward function evaluation) in the way we setup the experiment: In the previous experiment, we build POMDP with one value for (opponent) Type in the state space and then experiment by changing the reward function which should have been assigned for the other opponent types. In this experiment, we first make changes to the transcripts that are used to build the POMDP. In particular, we consider two setups here: Setup 1 where we (manually) change the opponent type in the first half of input transcripts where opponent type is Individualist and Setup 2 where we manually change the opponent type in the second half of transcripts where opponent type is Individualist, i.e., we introduce misclassification of opponent type in the input transcripts. For both the setups, we model two opponent types in state space i.e., value for Type in state space can be Individualist or Altruist. For both the setups we then perform three experiments: (a) Individualist correct i.e., no (manual) changes made to transcript and reward function used is the one that corresponds to when opponent type is individualist (b) Individualist Wrong as Altruists i.e., in the manually changed transcripts when Type is Altruist, POMDP continues to use Individualist reward since Altruist was introduced via a manual change (correct reward used when Type is Individualist) and (c) Individualist-Altruist correct (POMDP uses reward for Altruist when Type is Altruist).

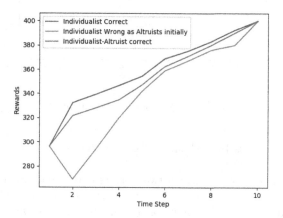

Fig. 3. Misclassification in the first half.

Figure 3 shows results for Setup 1, while Fig. 4 shows results for Setup 2. The x-axis for both the figures represent Timesteps while y-axis shows the EETR values. Both the figure shows that the orange line has the lowest EETR values due to the effect of type misclassification in input files as well as absence of correction in reward function, i.e., reward function used as if the misclassification was not present. The green line does better since, although misclassfication of opponent type was present, the reward function reflects this misclassification while the blue line performs best since there was no misclassfication in the input files (and hence no reward function correction is needed).

5   Experiment Results

5.1   Generation of Negotiation Data

We generate negotiation samples using the IAGO framework [16]. As described in [16], the negotiation samples are generated in IAGO when the player(s) negotiates with the IAGO agent. The IAGO agent can be part of a multi-issue bargaining task featuring four issues at five different levels. The agent utilizes the fixed list of utterances that the human may use, although it has its own set of responses. The agent attempts to gain the most value for itself in the negotiation by employing several human-negotiation techniques, such as appealing to

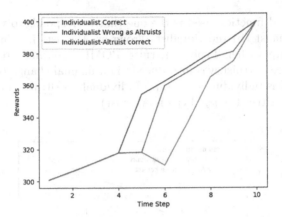

Fig. 4. Misclassification in the second half.

the principle of fairness by utilizing a fixed-pie lie. We will refer to the IAGO software agent as IAGOAgent.

We consider a scenario in which two players determine how to split a set of items between them. The items considered here are bars of gold, bars of iron, shipments of bananas and shipments of spices. The scenario was played out with the IAGOAgent and logs of the negotiation [16] were used to generate negotiation transcripts. The players had to negotiate with the IAGOAgent with a set time limit. Each of these players has different personality types. We used IAGOAgent versus the 4 personality types and generated a total of 30 transcripts for each of IAGOAgent-Individualist, IAGOAgent-Cooperator, IAGOAgent-Competitor and IAGOAgent-Altruist negotiation.

5.2 Opponent Personality Type Prediction

The key goal of our POMDP is to capture the personality type of the players and reason about them. In this experiment, we check whether it has indeed been successful in learning the behaviors using a belief tracking experiment [18]. To perform this experiment, we only use the POMDP belief update and ignore the rewards and optimal policy. We pre-specify the actions the POMDP agent is supposed to take as well as actions of the opponent player which are received as input directly from a dataset. Given this setup, we check if the POMDP can classify the opponent player type correctly using belief updates, i.e., if the type of other player is X, the sum of probabilities of all the states where type $=X$ would be the belief in player X. A high classification accuracy implies that the POMDP was able to learn the behaviors of the opponent correctly. In terms of setting up the experiment, the POMDP agent is trained using the actions which the IAGOAgent takes while the opponent player can be one of the four personality types namely Individualist, Cooperative, Competitive or Altruist (note that transcripts for each personality type are generated by IAGO via role play of each personality type during negotiation with the IAGOAgent).

- **Pair-wise** - In this experiment, we perform a pairwise comparison of the personality types, which leads to a total of 6 experimental settings, namely Individualist vs. Altruist, Individualist vs. Competitor, Individualist vs. Cooperator, Cooperator vs. Competitor, Cooperator vs. Altruist, Competitor vs. Altruist.

- **One vs. Rest Combined** - In this experiment, we perform a comparison of one personality type against a combination of the other three types, which leads to a total of 4 experimental settings, namely Individualist vs. Altruist & Competitor & Cooperator, Altruist vs. Competitor & Cooperator & Individualist, Competitor vs. Individualist & Cooperator & Altruist, Cooperator vs. Competitor & Altruist & Individualist.

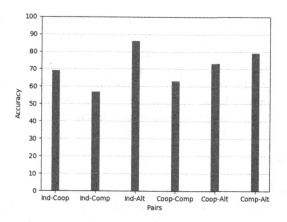

Fig. 5. Pair-wise comparison results.

For both the experiments, out of the 30 datasets generated for each personality type, we use 20 of them as training data and the other 10 datasets as test data. We use action pairs that are taken directly from the test data instead of allowing the POMDP to pick the optimal action. Figure 5 shows the classification accuracy values for the pair-wise comparison experiment. The X-axis indicates the pairs we have used as opponent players for testing, and Y-axis shows the accuracy in classification. Accuracy in classification refers to the number of times the POMDP belief update has correctly classified the opponent type.

In a Pair-wise tracking experiment, the POMDP initially believes that the opponent could be from any of the two personality types with a 0.5 belief; hence it is counted as a correct classification if the final belief at the end of the experiment is greater than 50%. Figure 5 shows that the Altruist type can be distinguished with high accuracy from the other types in a pair-wise comparison test with an accuracy greater than 70% in all the tests. Altruist is particularly distinguishable since they tend to maximize other agent's outcome rather than their own. The Cooperator type is second best with greater than 60% classification accuracy.

Individualists and Competitors fare worse (50% to 60%) since they have fairly similar characteristics (win vs. win big).

Table 1. Accuracy values

Personality type	Accuracy
Individualist	71%
Cooperator	75%
Competitor	65%
Altruist	81%

The second belief tracking experiment tries to distinguish between opponents where one opponent is of a single personality type while the other is a combination of 3 personality types. The goal of this experiment is to observe if we can get a better accuracy result than the pair-wise comparison. Table 1 shows the classification accuracy for this experiment. We observe that there is an improvement in classification accuracy for all the four types, possibly due to the consolidation of information into two categories instead of four categories earlier.

6 Conclusions

In this paper, we model an automated negotiation problem as a Partially Observable Markov Decision Process with a specific focus on personality types. We focus on four personality types, in particular, namely Individualist, Cooperator, Competitor, and Altruist. We generated 30 negotiation transcripts for each type for experimentation purposes using the IAGO software. For purposes of sanity check: (a) We compared our POMDP against two MDPs, MDP1, which does not model personality type and MDP2, which knows the personality type. We found $MDP1 \leq POMDP \leq MDP2$ in terms of performance (b) We introduced errors in POMDP model in multiple ways and found that they result in a decrease of reward obtained by the agent. Our classification accuracy results performed in two ways, i.e., Pair-wise and One vs. Rest, both show that we obtain good quality results, which showcase that transcripts containing personality-related information can indeed help to improve automated negotiation technology.

References

1. Baarslag, T., Hendrikx, M., Hindriks, K., Jonker, C.: Predicting the performance of opponent models in automated negotiation. In: 2013 IEEE/WIC/ACM International Joint Conferences on Web Intelligence (WI) and Intelligent Agent Technologies (IAT), vol. 2, pp. 59–66. IEEE (2013)

2. Baarslag, T., Hendrikx, M.J.C., Hindriks, K.V., Jonker, C.M.: Learning about the opponent in automated bilateral negotiation: a comprehensive survey of opponent modeling techniques. Auton. Agent. Multi-Agent Syst. **30**(5), 849–898 (2015). https://doi.org/10.1007/s10458-015-9309-1

3. Baarslag, T., Kaisers, M., Gerding, E., Jonker, C.M., Gratch, J.: When will negotiation agents be able to represent us? The challenges and opportunities for autonomous negotiators. In: International Joint Conferences on Artificial Intelligence (2017)

4. Buron, C.L.R., Guessoum, Z., Ductor, S.: MCTS-based automated negotiation agent. In: Baldoni, M., Dastani, M., Liao, B., Sakurai, Y., Zalila Wenkstern, R. (eds.) PRIMA 2019. LNCS (LNAI), vol. 11873, pp. 186–201. Springer, Cham (2019). https://doi.org/10.1007/978-3-030-33792-6_12

5. Camerer, C.F.: Behavioral game theory: predicting human behavior in strategic situations. In: Advances in Behavioral Economics, pp. 374–392 (2004)

6. Chen, S., Weiss, G.: OMAC: a discrete wavelet transformation based negotiation agent. In: Marsa-Maestre, I., Lopez-Carmona, M.A., Ito, T., Zhang, M., Bai, Q., Fujita, K. (eds.) Novel Insights in Agent-based Complex Automated Negotiation. SCI, vol. 535, pp. 187–196. Springer, Tokyo (2014). https://doi.org/10.1007/978-4-431-54758-7_13

7. De Jonge, D., Zhang, D.: Automated negotiations for general game playing. In: Proceedings of the 16th Conference on Autonomous Agents and MultiAgent Systems, pp. 371–379 (2017)

8. Fatima, S.S., Wooldridge, M., Jennings, N.R.: Multi-issue negotiation under time constraints. In: Proceedings of the First International Joint Conference on Autonomous Agents and Multiagent Systems: Part 1, pp. 143–150 (2002)

9. Hao, J., Leung, H.: CUHKAgent: an adaptive negotiation strategy for bilateral negotiations over multiple items. In: Marsa-Maestre, I., Lopez-Carmona, M.A., Ito, T., Zhang, M., Bai, Q., Fujita, K. (eds.) Novel Insights in Agent-based Complex Automated Negotiation. SCI, vol. 535, pp. 171–179. Springer, Tokyo (2014). https://doi.org/10.1007/978-4-431-54758-7_11

10. Ji, S., Zhang, C., Sim, K.-M., Leung, H.: A one-shot bargaining strategy for dealing with multifarious opponents. Appl. Intell. **40**(4), 557–574 (2013). https://doi.org/10.1007/s10489-013-0497-6

11. Kaelbling, L.P., Littman, M.L., Cassandra, A.R.: Planning and acting in partially observable stochastic domains. Artif. Intell. **101**(1–2), 99–134 (1998)

12. Kraus, S., Arkin, R.C.: Strategic Negotiation in Multiagent Environments. MIT Press, Cambridge (2001)

13. Kurniawati, H., Hsu, D., Lee, W.S.: SARSOP: efficient point-based POMDP planning by approximating optimally reachable belief spaces. In: Robotics: Science and systems, Zurich, Switzerland, vol. 2008 (2008)

14. Leahu, H., Kaisers, M., Baarslag, T.: Automated negotiation with Gaussian process-based utility models. In: Proceedings of the Twenty-eighth International Joint Conference on Artificial Intelligence, IJCAI, vol. 19 (2019)

15. Leahu, H., Kaisers, M., Baarslag, T.: Preference learning in automated negotiation using Gaussian uncertainty models. In: Proceedings of the 18th International Conference on Autonomous Agents and MultiAgent Systems, pp. 2087–2089. International Foundation for Autonomous Agents and Multiagent Systems (2019)

16. Mell, J., Gratch, J.: IAGO: interactive arbitration guide online. In: AAMAS, pp. 1510–1512 (2016)

17. Mell, J., Gratch, J.: Grumpy & Pinocchio: answering human-agent negotiation questions through realistic agent design. In: Proceedings of the 16th Conference on Autonomous Agents and Multiagent Systems, pp. 401–409 (2017)

18. Paruchuri, P., Chakraborty, N., Gordon, G., Sycara, K., Brett, J., Adair, W.: Intercultural opponent behavior modeling in a POMDP based automated negotiating agent. In: Sycara, K., Gelfand, M., Abbe, A. (eds.) Models for Intercultural Collaboration and Negotiation. AGDN, vol. 6, pp. 165–182. Springer, Dordrecht (2013). https://doi.org/10.1007/978-94-007-5574-1_9

19. Paruchuri, P., Chakraborty, N., Zivan, R., Sycara, K., Dudik, M., Gordon, G.: POMDP based negotiation modeling. In: Proceedings of the IJCAI Workshop on Modeling Intercultural Collaboration and Negotiation (MICON) (2009)

20. Pineau, J., Gordon, G., Thrun, S., et al.: Point-based value iteration: An anytime algorithm for POMDPs. In: IJCAI, vol. 3, pp. 1025–1032 (2003)

21. Pynadath, D.V., Marsella, S.C.: Psychsim: modeling theory of mind with decision-theoretic agents. In: IJCAI, vol. 5, pp. 1181–1186 (2005)

22. Rosenfeld, A., Kraus, S.: Strategical argumentative agent for human persuasion. In: Proceedings of the Twenty-Second European Conference on Artificial Intelligence, pp. 320–328. IOS Press (2016)

23. Salnikov, M.: Text clustering with k-means and TF-IDF. https://medium.com/@MSalnikov/text-clustering-with-k-means-and-tf-idf-f099bcf95183 (2018). Accessed Aug 2019

24. Santos, R., Marreiros, G., Ramos, C., Neves, J., Bulas-Cruz, J.: Using personality types to support argumentation. In: McBurney, P., Rahwan, I., Parsons, S., Maudet, N. (eds.) ArgMAS 2009. LNCS (LNAI), vol. 6057, pp. 292–304. Springer, Heidelberg (2010). https://doi.org/10.1007/978-3-642-12805-9_17

25. Shonk, K.: Understanding different negotiation styles. https://www.pon.harvard.edu/daily/negotiation-skills-daily/understanding-different-negotiation-styles/ (2019). Accessed Oct 2019

26. Yu, C., Ren, F., Zhang, M.: An adaptive bilateral negotiation model based on Bayesian learning. In: Ito, T., Zhang, M., Robu, V., Matsuo, T. (eds.) Complex Automated Negotiations: Theories, Models, and Software Competitions. SCI, vol. 435, pp. 75–93. Springer, Berlin (2013). https://doi.org/10.1007/978-3-642-30737-9_5

27. Zeng, D., Sycara, K.: Bayesian learning in negotiation. Int. J. Hum Comput Stud. **48**(1), 125–141 (1998)

Decision Rule Aggregation Approach to Support Group Decision Making

Inès Saad[1,2(✉)] and Salem Chakhar[3]

[1] MIS, University of Picardie Jules Verne, Amiens, France
ines.saad@u-picardie.fr
[2] Amiens Business School, Amiens, France
[3] Portsmouth Business School and CORL, University of Portsmouth,
Portsmouth, UK
salem.chakhar@port.ac.uk

Abstract. The Dominance-based Rough Set Approach (DRSA) is an innovative preference learning approach. It takes as input a set of objects (learning set) described with respect to a collection of condition and decision attributes. It generates a set of if-then decision rules. Initial versions of dominance based rough set approximation methods assume a single decision maker. Furthermore, the proposed extensions to group decision making mainly use an input oriented aggregation strategy, which requires a high level of agreement between the decision makers. In this paper, we propose an output oriented aggregation strategy to coherently combine different sets of decision rules obtained from different decision makers. The proposed aggregation algorithm is illustrated by using real-world data relative to a business school admission where two decision makers are involved. Results show that aggregation algorithm is able to reproduce the individual assignments of students with a very limited preferential information loss.

Keywords: Rough set approximation · Decision rule · Group decision making · Rules aggregation · Output aggregation strategy

1 Introduction

The Dominance-based Rough Set Approach (DRSA) [7] is an extension of the Rough Sets Theory [10] intended to deal with multicriteria sorting problems. The DRSA takes a set of assignment examples (learning set) and generates a collection of if-then decision rules as output. The conventional DRSA assumes a single decision maker while several real-world decision problems need to take into account the presence of multiple decision makers. Different group decision making extensions to DRSA have been proposed in the literature, including [1–6, 8, 11, 12].

For instance, the authors in [1] and [8] extend the concepts of the DRSA to deal with decision tables having multiple decision attributes, thus allowing

© Springer Nature Switzerland AG 2020
D. C. Morais et al. (Eds.): GDN 2020, LNBIP 388, pp. 163–176, 2020.
https://doi.org/10.1007/978-3-030-48641-9_12

comprehensive collective decision rules to be generated. In [4] we introduced an aggregation algorithm, based on the majority principle and supporting the veto effect, allowing consensual decision rules to be inferred. A more advanced version of the aggregation algorithm of [4] is proposed in [3]. In [5,6], the authors use the Dempster-Shafer theory of evidence to combine individual rules provided by the DRSA.

However, all these approaches rely on an input oriented aggregation strategy, which requires a high level of agreement between the decision makers. In this paper, we propose an output oriented aggregation strategy to coherently combine different sets of decision rules obtained from different decision makers. The proposed aggregation algorithm is illustrated by using real-world data relative to business school admission where two decision makers are involved. Results show that aggregation algorithm is able to reproduce the individual assignments of students with a very low preferential information loss.

The rest of the paper is structured as follows. Section 2 sets the background. Section 3 deals with rules matching and overleaping. Section 4 details the aggregation algorithm. Section 5 provides an illustrative application. Section 6 concludes the paper.

2   Background

2.1   Notations and Basic Assumptions

Information about decision objects is often represented in terms of an *information table* where rows correspond to *objects* and columns to *attributes*. The information table S is a 4-tuple $<U, Q, V, f>$ where: U is a finite set of objects, Q is a finite set of attributes, $V = \bigcup_{q \in Q} V_q$, where V_q is a domain of the attribute q, and $f : U \times Q \rightarrow V$ an *information function* defined such that $f(x, q) \in V_q, \forall q \in Q, \forall x \in U$. The set of attributes Q is often divided into a sub-set C of *condition attributes* and a sub-set D of *decision attributes*. In this case, S is called *decision table*.

The domain of condition attributes is supposed to be ordered in decreasing or increasing preference. Such attributes are called *criteria*. We assume that the preference is increasing with value of $f(\cdot, q)$ for every $q \in C$. We also assume that the set of decision attributes D is a singleton $\{d\}$. Decision attribute d makes a partition of U into a finite number of decision classes $\mathbf{Cl} = \{Cl_t, t \in T\}$, $T = \{0, \cdots, n\}$, such that each $x \in U$ belongs to one and only one class in $\mathbf{Cl}$. Further, we assume that the classes are preference-ordered, i.e. for all $r, s \in T$, such that $r > s$, the objects from Cl_r are preferred to the objects from Cl_s.

2.2   Rough Approximation

In DRSA the represented knowledge is a collection of *upward unions* $Cl_t^{\geq}$ and *downward unions* $Cl_t^{\leq}$ of classes defined as follows: $Cl_t^{\geq} = \cup_{s \geq t} Cl_s$ and $Cl_t^{\leq} = \cup_{s \leq t} Cl_s$. The assertion "$x \in Cl_t^{\geq}$" means that "$x$ belongs to at least class Cl_t"

while assertion "$x \in Cl_t^{\leq}$" means that "x belongs to at most class Cl_t". The basic idea of DRSA is to replace the indiscernibility relation used in the conventional Rough Set Theory with a dominance relation. Let $P \subseteq C$ be a subset of condition attributes. The *dominance relation* Δ_P associated with P is defined for each pair of objects x and y as follows:

$$x \Delta_P y \Leftrightarrow f(x, q) \succeq f(y, q), \forall q \in P. \tag{1}$$

In the definition above, the symbol "$\succeq$" should be replaced with "$\preceq$" for condition attributes which are ordered according to decreasing preferences. To each object $x \in U$, we associate two sets: (i) the *P-dominating set* $\Delta_P^+(x) = \{y \in U : y\Delta_P x\}$ containing the objects that dominate x, and (ii) the *P-dominated set* $\Delta_P^-(x) = \{y \in U : x\Delta_P y\}$ containing the objects dominated by x.

Then, the *P*-lower and *P*-upper approximations of $Cl_t^{\geq}$ with respect to P are defined as follows:

- $\underline{P}(Cl_t^{\geq}) = \{x \in U : \Delta_P^+(x) \subseteq Cl_t^{\geq}\}$,
- $\bar{P}(Cl_t^{\geq}) = \{x \in U : \Delta_P^-(x) \cap Cl_t^{\geq} \neq \emptyset\}$.

Analogously, the *P*-lower and *P*-upper approximations of $Cl_t^{\leq}$ with respect to *P* are defined as follows:

- $\underline{P}(Cl_t^{\leq}) = \{x \in U : \Delta_P^-(x) \subseteq Cl_t^{\leq}\}$,
- $\bar{P}(Cl_t^{\leq}) = \{x \in U : \Delta_P^+(x) \cap Cl_t^{\leq} \neq \emptyset\}$.

The lower approximations group the objects which certainly belong to class unions $Cl_t^{\geq}$ (resp. $Cl_t^{\leq}$). The upper approximations group the objects which could belong to $Cl_t^{\geq}$ (resp. $Cl_t^{\leq}$).

The *P*-boundaries of $Cl_t^{\geq}$ and $Cl_t^{\leq}$ are defined as follows:

- $Bn_P(Cl_t^{\geq}) = \bar{P}(Cl_t^{\geq}) - \underline{P}(Cl_t^{\geq})$,
- $Bn_P(Cl_t^{\leq}) = \bar{P}(Cl_t^{\leq}) - \underline{P}(Cl_t^{\leq})$.

The boundaries group objects that can neither be ruled in nor out as members of class Cl_t.

2.3 Decision Rules

The approximations of upward and downward unions of classes can serve to induce a set of if-then decision rules relating condition and decision attributes. There are five basic types of decision rules:

- *Certain decision rules.* These rules are generated from the lower approximation of the union of classes $Cl_t^{\leq}$ or $Cl_t^{\geq}$. A decision rule from this type has one of the following structures:
  - **Type 1**: **if** $f(x, q_1) \leq r_1 \wedge \cdots \wedge f(x, q_m) \leq r_m$ **then** $x \in Cl_t^{\leq}$
  - **Type 2**: **if** $f(x, q_1) \geq r_1 \wedge \cdots \wedge f(x, q_m) \geq r_m$ **then** $x \in Cl_t^{\geq}$

where $(r_1, \cdots, r_m) \in (V_{q_1} \times \cdots \times V_{q_m})$.

- *Possible decision rules.* These rules are generated from the upper approximation of the union of classes $Cl_t^{\leq}$ or $Cl_t^{\geq}$. A decision rule from this type has one of the following structures:
  - **Type 3: if** $f(x, q_1) \leq r_1 \wedge \cdots \wedge f(x, q_m) \leq r_m$ **then** x could belong to $Cl_t^{\leq}$
  - **Type 4: if** $f(x, q_1) \geq r_1 \wedge \cdots \wedge f(x, q_m) \geq r_m$ **then** x could belong to $Cl_t^{\geq}$

  where $(r_1, \cdots, r_m) \in (V_{q_1} \times \cdots \times V_{q_m})$.

- *Approximate rules.* These rules are generated from the boundaries. A decision rule from this type has the following structure:
  - **Type 5: if** $f(x, q_1) \leq r_1 \wedge \cdots \wedge f(x, q_m) \leq r_m \wedge f(x, q_{m+1}) \leq r_{m+1} \wedge \cdots \wedge f(x, r_p) \leq r_p$ **then** $x \in Cl_s \cup Cl_{s+1} \cup \cdots \cup Cl_t$

  where $(r_1, \cdots, r_p) \in (V_{q_1} \times \cdots \times V_{q_p})$.

Only the two first types are considered in the rest of this paper.

The most popular rule induction algorithm for DRSA is DOMLEM [9], which generates a minimal set of rules.

3   Decision Rules Matching and Overlapping

3.1   Basic Definitions

A decision rule R is defined as a collection of elementary conditions and a conclusion. Let $R.C$ denotes the set of conditions of rule R and $R.C_i$ denote each member of this set. Let $R.N$ denote the cardinality of this set. Let $R.D$ denotes the conclusion associated with rule R. Each decision rule R is characterized by its type $R.T$.

An elementary condition C_i is defined by an attribute Q, an operator O and a right-hand member H. Let $C_i.Q$, $C_i.O$ and $C_i.H$ denote these three elements.

3.2   Conditions Matching

Definition 1 (Conditions equality). *Let C_k and C_l be two conditions of the same type. Then, C_k is equal to C_l (denoted $C_k = C_l$) iff:*

$$\left.\begin{array}{l} C_k.Q = C_l.Q \\ C_k.O = C_l.O \\ C_k.H = C_l.H \end{array}\right\} \Rightarrow (C_k = C_l)$$

Definition 2 (Type 1 conditions inclusion). *Let C_k and C_l be two conditions of Type 1 decision rules. Then, C_k is included in C_l (denoted $C_k \subseteq C_l$) iff:*

$$\left.\begin{array}{l} C_k.Q = C_l.Q \\ C_k.O = C_l.O \\ C_k.H \preceq C_l.H \end{array}\right\} \Rightarrow (C_k \subseteq C_l)$$

Definition 3 (Type 2 conditions inclusion). *Let C_k and C_l be two conditions of Type 2 decision rules. Then, C_k includes C_l (denoted $C_k \supseteq C_l$) iff:*

$$\left\{ \begin{array}{l} C_k.Q = C_l.Q \\ C_k.O = C_l.O \\ C_k.H \succeq C_l.H \end{array} \right\} \Rightarrow (C_k \supseteq C_l)$$

3.3 Decision Rules Matching

The equality between two decision rules is defined as follows.

Definition 4 (Decision rules equality). *Let R_i and R_j be two decision rules of the same type. Then, R_i is equal to R_j (denoted $R_i = R_j$) iff:*

$$\left\{ \begin{array}{l} R_i.T = R_j.T \\ \forall k \exists l (R_i.C_k = R_j.C_l) \quad 1 \leq k \leq R_i.N \\ \forall m \exists n (R_j.C_m = R_i.C_n) \quad 1 \leq m \leq R_j.N \\ R_i.D = R_j.D \end{array} \right\} \Rightarrow (R_i = R_j)$$

Definition 5 (Decision rules Type 1 full inclusion). *Let R_i and R_j be two decision rules of Type 1. Then, R_i is fully included in R_j (denoted $R_i \subseteq R_j$) iff:*

$$\left\{ \begin{array}{l} R_i.T = R_j.T = Type\ 1 \\ \forall k \exists l (R_i.C_k \subseteq R_j.C_l) \quad 1 \leq k \leq R_i.N \\ \forall m \exists n (R_i.C_n \subseteq R_j.C_m) \quad 1 \leq m \leq R_i.N \\ R_i.D \preceq R_j.D \end{array} \right\} \Rightarrow (R_i \subseteq R_j)$$

This definition implicity ensures that rules R_i and R_j must have the same cardinality, i.e., $R_i.N = R_j.N$.

Definition 6 (Decision rules Type 1 Partial inclusion). *Let R_i and R_j be two decision rules of Type 1. Then, R_i is partially included in R_j (denoted $R_i \subset R_j$) iff:*

$$\left\{ \begin{array}{l} R_i.T = R_j.T = Type\ 1 \\ \forall k \exists l (R_i.C_k \subseteq R_j.C_l) \quad 1 \leq k \leq R_i.N \\ R_i.D \preceq R_j.D \\ R_i.N < R_j.N \end{array} \right\} \Rightarrow (R_i \subset R_j)$$

The last condition (i.e., $R_i.N < R_j.N$) in this definition ensures that the cardinality of rule R_i must be strictly less that the one of R_j.

Definition 7 (Decision rules Type 2 full inclusion). *Let R_i and R_j be two decision rules of Type 2. Then, R_i is fully included in R_j (denoted $R_i \supseteq R_j$) iff:*

$$\left\{ \begin{array}{l} R_i.T = R_j.T = Type\ 2 \\ \forall k \exists l (R_i.C_k \supseteq R_j.C_l) \quad 1 \leq k \leq R_i.N \\ \forall m \exists n (R_i.C_n \subseteq R_j.C_m) \quad 1 \leq m \leq R_i.N \\ R_i.D \succeq R_j.D \end{array} \right\} \Rightarrow (R_i \supseteq R_j)$$

This definition implicity ensures that rules R_i and R_j must have the same cardinality, i.e., $R_i.N = R_j.N$.

Definition 8 (Decision rules Type 2 partial inclusion). *Let R_i and R_j be two decision rules of Type 2. Then, R_i is partially included in R_j (denoted $R_i \supset R_j$) iff:*

$$\left.\begin{cases} R_i.T = R_j.T = Type\ 2 \\ \forall k \exists l (R_i.C_k \supseteq R_j.C_l) \quad 1 \le k \le R_i.N \\ R_i.D \succeq R_j.D \\ R_i.N < R_j.N \end{cases}\right\} \Rightarrow (R_i \supset R_j)$$

The last condition (i.e., $R_i.N < R_j.N$) in this definition ensures that the cardinality of rule R_i must be strictly less that the one of R_j.

3.4 Overlapping Decision Rules

Let R_i be a Type 1 decision rule and R_j be a Type 2 decision rule. Although these rules are of different types, they may or not share some parts of their conditions and/or decisions. Four basic cases can be distinguished: (i) R_i and R_j are fully disjoint; (ii) R_i and R_j have overlapped conditions but their decisions are disjoint; (iii) R_i and R_j have overlapped decisions but their conditions are disjoint; and (iv) R_i and R_j have overlapped conditions and overlapped decisions.

Definition 9 (Decision rules with overlapped conditions). *Let R_i be a Type 1 decision rule and R_j be a Type 2 decision rule. Decision rules R_i and R_j have overlapped conditions, denoted $(R_i.C \cap R_j.C) \ne \emptyset$, iff:*

$$\left.\begin{cases} \forall C_k \in R_i.C, \exists C_l \in R_j.C(C_k.Q = C_l.Q) \wedge \\ (C_k.O \ne C_l.O) \wedge (C_k.H \ge C_l.H)1 \le k \le R_i.N \\ \forall C_m \in R_j.C, \exists C_n \in R_i.C(C_n.Q = C_m.Q) \wedge \\ (C_n.O \ne C_m.O) \wedge (C_m.H \le C_n.H)1 \le m \le R_j.N \end{cases}\right\} \Rightarrow (R_i.C \cap R_j.C) \ne \emptyset$$

Definition 10 (Decision rules with overlapped decisions). *Let R_i be a Type 1 decision rule and R_j be a Type 2 decision rule. Decision rules R_i and R_j have overlapped decisions, denoted $(R_i.D \cap R_j.D) \ne \emptyset$, iff only if: $R_i.D \preceq R_j.D$.*

Definition 11 (Fully overlapped decision rules). *Let R_i be a Type 1 decision rule and R_j be a Type 2 decision rule. Decision rules R_i and R_j are fully overlapped, denoted $(R_i.C \cap R_j.C) \ne \emptyset$, iff:*

$$\left.\begin{cases} \forall C_k \in R_i.C, \exists C_l \in R_j.C(C_k.Q = C_l.Q) \wedge \\ \quad (C_k.O \ne C_l.O) \wedge (C_k.H \ge C_l.H) \\ \forall C_m \in R_j.C, \exists C_n \in R_i.C(C_n.Q = C_m.Q) \wedge \\ \quad (C_n.O \ne C_m.O) \wedge (C_m.H \le C_n.H) \\ R_i.D \succeq R_j.D \end{cases}\right\} \Rightarrow (R_i \cap R_j) \ne \emptyset$$

Definition 12 (Disjoint decision rules). *Let R_i be a Type 1 decision rule and R_j be a Type 2 decision rule. Decision rules R_i and R_j are disjoint, denoted $(R_i \cap R_j) = \emptyset$, iff:*

$$\left\{ \begin{array}{l} \forall C_k \in R_i.C, \exists C_l \in R_j(C_k.Q = C_l.Q) \wedge \\ \quad (C_k.O \neq C_l.O) \wedge (C_k.H < C_l.H) \\ \forall C_l \in R_j.C, \exists C_k \in R_i(C_l.Q = C_k.Q) \wedge \\ \quad (C_l.O \neq C_k.O) \wedge (C_l.H > C_k.H) \\ R_i.D \prec R_j.R \end{array} \right\} \Rightarrow (R_i \cap R_j) = \emptyset$$

4 Decision Rules Aggregation

Let $H = \{1, \cdots, i, \cdots, h\}$ be a set of decision makers and Π_i the set decision rules obtained by decision maker $i \in H$. Let Π be the union of all decision rules of the h decision makers: $\Pi = \cup_{i=1}^{h} \Pi_i$.

The aggregation algorithm contains two steps: (i) transformation of overlapping rules, and (ii) elimination of redundant decision rules. We should mention that steps (i) and (ii) may be inverted without affecting the final result. However, in this paper, we maintain the order given above for several reasons. First, the other solution (i.e. proceeding by computing the minimal cover and then transformation of overlapping rules) requires an additional step to compute the minimal cover after the transformation operation. Indeed, the latter may lead to new redundant rules. Second, as a consequence of the first point, the computing time will automatically increase. The only shortcoming of the solution adopted in this paper is that in step (i) both redundant and non-redundant rules are considered. This may have minor effects on the overall computing time.

The aggregation algorithm takes the set Π of all decision rules and generates a minimal set of non-redundant decision rules.

4.1 Step 1: Transformation of Overlapping Decision Rules

Case 1. R_i and R_j Are Disjoint Decision Rules. This situation is graphically illustrated by Fig. 1. In this figure, we assumed that all conditions attributes have the same scale and that $1 < \alpha < \beta < n$. As it is shown in Fig. 1, the constraints defined by the conditions of rules R_i and R_j are totaly disjoint. For instance, condition C_2 says that $f(x, A_s) \leq r_\beta$ and condition $C_{2'}$ says that $f(x, A_s) \geq r_n$. It is easy to see that there is no any intersection between the two constraints as defined by C_2 and $C_{2'}$. The same remark holds for the other conditions and for the decision of R_i and R_j.

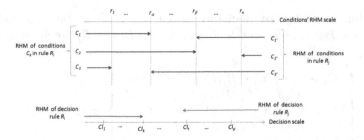

Fig. 1. Schematic representation of disjoint decision rules.

In this situation, there is no overlap between decision rules R_i and R_j and it is reasonable to maintain both of them (if they are not overlapped by other rules).

Case 2. R_i and R_j Have Overlapped Conditions and Decisions This situation is graphically illustrated by Fig. 2. In this figure, we assumed that all conditions attributes have the same scale and that $1 < \alpha < \beta < n$. As it is shown in Fig. 2, the constraints defined by the conditions of rules R_i and R_j overlap. For instance, condition C_2 says that $f(x, A_s) \leq r_n$ and condition $C_{2'}$ says that $f(x, A_s) \geq r_\beta$. It is easy to see that there is an intersection between the two constraints defined by C_2 and $C_{2'}$ (since $\beta < n$). The same remark holds for the other conditions. The same remark holds for decisions of R_i and R_j.

To reduce the interval-based assignments of decision objects, we propose to replace rules R_i and R_j by three fully disjoint decision rules as follows:

- R_a: with the same structure and type as rule R_i but the Right Hand Side (RHS) of conditions and the decision are those of rule R_j;
- R_b: with the same structure and type as rule R_j but the RHS of conditions and the decision are those of rule R_i;
- R_c: the RHS of conditions are of the form $[R_j.C_k.H, R_i.C_k.H]$ and the decision if of the form $[R_j.D, R_i.D]$.

The last decision rule is of composite type since RHS of the conditions and the decision are interval-based.

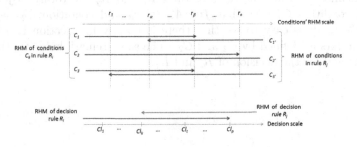

Fig. 2. Schematic representation of full overlap of decision rules

Case 3. R_i and R_j Have Overlapped Conditions This situation is graphically illustrated by Fig. 3. In this figure, we assume that all conditions attributes have the same scale and that $1 < \alpha < \beta < n$. As it is shown in Fig. 3, the constraints defined by the conditions of rules R_i and R_j overlap but the decisions do not. For instance, condition C_2 says that $f(x, A_s) \leq r_n$ are condition $C_{2'}$ says that $f(x, A_s) \geq r_\beta$. It is easy to see that there is any intersection between the two constraints defined by C_2 and $C_{2'}$ (since $\beta < n$). The same remark holds for the other conditions. On the contrary, the decision parts of rules R_i and R_j are totaly disjoint.

To reduce the interval-based assignments of decision objects, we propose to replace rules R_i and R_j with two certain and more precise decision rules as follows:

- R_a: with the same structure and type as rule R_i but the RHS of conditions and the decision are those of rule R_j;
- R_b: with the same structure and type as rule R_j but the RHS of conditions and the decision are those of rule R_i;

We mention that we may identify a third situation concerning the assignment of objects where the RHS of the different conditions are in the range $[r_\alpha, r_\beta]$ (see Fig. 3). In this case, there is a contradiction with the initial decision:

- by R_i, objects with RHS in $[r_\alpha, r_\beta]$ should be assigned to $Cl_{\overline{k}}^{\leq}$, and
- by R_i, objects with RHS in $[r_\alpha, r_\beta]$ should be assigned to $Cl_{\overline{t}}^{\geq}$,

Since $k < t$, we should assign objects either to $Cl_{\overline{k}}^{\leq}$ or $Cl_{\overline{t}}^{\geq}$. However, to avoid conflict assignments, we opted out not to include an additional rule as in the previous case.

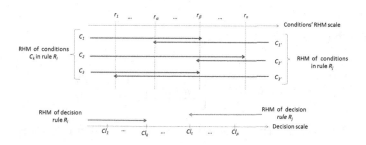

Fig. 3. Schematic representation of conditions overlap

Case 4. R_i and R_j Have Overlapped Decisions This situation is graphically illustrated by Fig. 4. In this figure, we assumed that all conditions attributes have the same scale and that $1 < \alpha < \beta < n$. As shown in Fig. 4 the constraints defined by the conditions of rules R_i and R_j are totaly disjoint while decisions overlap.

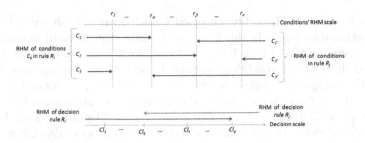

Fig. 4. Schematic representation of decisions overlap

To reduce the interval-based assignments of decision objects, we propose to replace rules R_i and R_j with two certain and more precise decision rules as follows:

- R_a: like rule R_i but the decision is as rule R_j;
- R_b: like rule R_j but the decision is as rule R_i;

4.2   Step 2: Elimination of Redundant Decision Rules

The objective of this step is to eliminate (i) redundant decision rules; and (ii) rules fully included in other rules. In the second case, two options are possible: either we remove the more general rule or the less general rule. Both solutions may lead to preferential information loss. To minimize the loss of preferential information, we can rely on some measures. Let R_a and R_b be two redundant decision rules. Let $[[R_a]]$ and $[[R_b]]$ be the sets of decision objects supporting decision rules R_a and R_b, respectively. Then, we define the following two measures:

- Information loss:

$$IL(R_a, R_b) = \begin{cases} 0, & \text{if } R_a \subseteq R_b, \\ \frac{[[R_b]] \setminus [[R_a]]}{[[R_b]]}, & \text{otherwise.} \end{cases} \qquad (2)$$

$IL(R_a, R_b)$ measures the information loss when decision rule R_b is removed.
- Precision loss:

$$PL(R_a, R_b) = 1 - IL(R_a, R_b). \qquad (3)$$

These two measures vary in different directions and can be used to decide which of decision rules R_a and R_b should be removed. It consists of a tradeoff between information loss and precision loss.

5 Application

To partially illustrate the proposed algorithm, we consider a real-world data relative to a business school admission where two decision makers (designed by DM1 and DM2 in the rest of the paper) are involved. The learning set is composed of 175 objects (students in this case). A randomly selected extract from the learning set is given in Table 1. In this table, the decisions 'A' and 'R' stand for 'accepted' and 'rejected', respectively. The comparison of the individual assignments shows that the decision makers disagree on 40 (22.86%) students.

Table 1. An extract from the learning set.

ID	French (fr)	Logic (lo)	General Culture	Collective interview (co)	Face to Face Meeting (ff)	Individual interview (in)	English (en)	Decision DM1 (dec1)	Decision DM2 (dec2)
135	16.8	10.6	6.6	10	Yes	13.5	8	A	A
162	9.6	10.6	6.6	14.83	Yes	9.5	6.5	R	R
161	14.4	8	10.8	11	Yes	5.6	15	R	A
73	12.8	13.4	6.6	14.33	Yes	14.8	12.5	A	A
64	13.6	20	10.8	9.83	Yes	15.6	10	A	A
110	14.4	10.6	3.4	9	Yes	14.9	14	A	R
159	14.4	13.4	9.2	6.83	Yes	11.8	6	R	A
89	9.6	10.6	5.8	9.83	Yes	18	11.5	A	R
15	12.8	9.4	11.6	?	No	18	15	A	A
1	16.8	12	10.8	18	Yes	17.4	19	A	A
117	15.2	9.4	11.6	11.33	Yes	12.9	10	A	A
53	10.4	9.4	5.8	14	Yes	17.4	14	A	R
144	9.6	12.6	15	?	No	11.3	7.25	A	A
45	13.6	17.4	6.6	18	Yes	13.5	12.5	A	A
80	13.6	14.6	5.8	18	Yes	14.1	8	A	A
41	13.6	16	10.8	15.33	Yes	12.7	16	A	A
147	10	10.6	8.4	?	No	13.5	6	A	R
160	14.4	8	7.6	17	Yes	8.5	6	R	A
84	16.8	16	2.6	9.33	Yes	15.6	12	A	A
123	17.6	12	5	11	Yes	12.5	11	A	A
33	12.8	13.4	9.2	13.67	Yes	18	12	A	A
82	14.4	10.6	7.6	18	Yes	14.7	7	A	A
14	15.2	15	13.4	?	No	16.6	14	A	A
168	13.6	10.6	5	4	Yes	12.7	6	R	R
104	15.2	14.6	13.4	13	Yes	10.7	11.5	A	A

We then applied the DRSA two times to approximate this learning set using the assignments given by DM1 and DM2. The application of rule induction algorithm DOMLEM on the obtained approximations leads to two collections of decision rules, which are given Table 2 (for DM1) and Table 3 (for DM2).

This illustrative example uses only two decision classes. Accordingly, there is no overlap between decision rules of different types. Then, only the second step will applied to aggregate the decision rules. A careful examination of Table 2 and Table 3 shows that there are three cases of redundancy: (i) Rule 1.9 and Rule 2.18; (ii) Rule 1.12 and Rule 2.11; and (iii) Rule 1.13 and Rule 2.16. The result

Table 2. Decision rules relative to DM1.

#	Rule description
1.1	If (in ≤ 7.5) then (dec1 = R)
1.2	If (en ≤ 1.5) then (dec1 = R)
1.3	If (en ≤ 4) & (gc ≤ 5) then (dec1 = R)
1.4	If (en ≤ 6.5) & (in ≤ 11.8) then (dec1 = R)
1.5	If (lo ≤ 6.6) & (gc ≤ 4.2) then (dec1 = R)
1.6	If (in ≤ 9) & (co ≤ 6) then (dec1 = R)
1.7	If (gc ≤ 5) & (en ≤ 6) & (in≤ 12.7) then (dec1 = R)
1.8	If (in ≥ 15.67) then (dec1 = A)
1.9	If (lo ≥ 20) then (dec1 = A)
1.10	If (co ≥ 15.6) & (lo ≥ 17.4) then (dec1 = A)
1.11	If (en ≥ 8.5) & (co ≥ 10.83) & (lo ≥ 14) then (dec1 = A)
1.12	If (gc ≥ 12) then (dec1 = A)
1.13	If (fr ≥ 16.8) then (dec1 = A)
1.14	If (in ≥ 9.2) & (en ≥ 4.56) & (ff ≥ 1) & (co ≥ 16) then (dec1 = A)
1.15	If (in ≥ 11.6) & (lo ≥ 7.4) & (en ≥ 4.56) then (dec1 = A)
1.16	If (en ≥ 10.75) & (gc ≥ 8.6) & (ff ≥ 1) & (co ≥ 10) & (en ≥ 12) & (in ≥ 9.2) then (dec1 = A)
1.17	If (lo ≥ 13.4) & (en ≥10.75) & (fr ≥ 13.6) & (gc ≥ 7.6) & (in ≥ 9.83) then (dec1 = A)

Table 3. Decision rules relative to DM2.

#	Rule description
2.1	If (fr ≤ 6.6) then (dec2 =R)
2.2	If (lo ≤ 5.4) & (gc ≤ 8.4) then (dec2 =R)
2.3	If (fr ≤ 10.8) & (gc ≤ 8.4) & (lo ≤ 14.6) then (dec2 =R)
2.4	If (lo ≤ 8.0) & (en ≤ 6.0) & (in ≤ 4.0) then (dec2 =R)
2.5	If (gc ≤ 5.0) & (lo ≤ 10.6) & (fr ≤ 14.4) then (dec2 =R)
2.6	If (lo ≤ 8.0) & (en ≤ 5.5) then (dec2 =R)
2.7	If (gc ≤ 5.8) & (en ≤ 9.5) & (lo ≤ 12.0) & (in ≤ 14.8) & (fr ≤ 12.8) then (dec2 =R)
2.8	If (lo ≤ 9.4) & (gc ≤ 6.6) & (en ≤ 11.0) & (in ≤ 13.1) then (dec2 =R)
2.9	If (lo ≤ 8.0) & (gc ≤ 5.8) & (en ≤ 9.5) & (fr ≤ 16.0) then (dec2 =R)
2.10	If (gc ≤ 5.8) & (lo ≤ 10.6) & (fr ≤ 12.8) then (dec2 =R)
2.11	If (gc ≥ 9.6) then (dec2 = A)
2.12	If (lo ≥ 15.2) & (en ≥ 16.5) then (dec2 = A)
2.13	If (fr ≥ 14.4) & (facetoface ≥ 1) & (in ≥ 18.0) then (dec2 = A)
2.14	If (lo ≥ 13.4) & (fr ≥ 15.2) then (dec2 = A)
2.15	If (fr ≥ 12.0) & (ff ≥ 1) & (gc ≥ 5.8) & (lo ≥ 6.6) & (in ≥ 8.0) then (dec2 = A)
2.16	If (fr ≥ 17.6) then (dec2 = A)
2.17	If (lo ≥ 10.6) & (in ≥ 18.0) & (gc ≥ 5.8) then (dec2 = A)
2.18	If (lo ≥ 17.4) then (dec2 = A)
2.19	If (en ≥ 14.0) & (fr ≥ 13.6) & (in ≥ 16.2) then (dec2 = A)
2.20	If (lo ≥ 13.4) & (in ≥ 11.67) & (ff ≥ 1) & (in ≥ 15.67) then (dec2 = A)
2.21	If (en ≥ 14.5) & (gc ≥ 8.2) then (dec2 = A)
2.22	If (lo ≥ 16.0) & (in ≥ 16.4) then (dec2 = A)
2.23	If (gc ≥ 9.2) & (lo ≥ 13.4) then (dec2 = A)

of the application of Equations (1) and (2) on these pairs of decision rules is summarized in Table 4. Based on these results and to reduce information loss, decision rules 1.18, 2.11 and 1.13 should be removed.

Table 4. Information and precision loss.

Case	Rule to remove	IL	PL
1	1.9	0	1
	1.18	0.923	0.077
2	1.12	0	1
	2.11	0.633	0.367
3	1.13	0.245	0.755
	2.16	0	1

We then applied the remaining decision rules to classify all the students. Results show that the obtained collective assignments match with the initial assignments of DM1 for about 96.2% of students and with the initial assignments of DM2 for about 92.3% of students. Thus, DM1 and DM2 need to discuss only a very limited number of conflicting situations (instead on 40 conflicting situations initially).

6   Conclusion

We proposed an output oriented aggregation strategy to coherently combine different sets of decision rules obtained from different decision makers. The proposed aggregation algorithm is illustrated by using real-world data relative to business school admission. An important aspect of the proposed approach is that the consensus between decision makers [13] is computed using objective preference information. In the future, we intend first to apply the proposed aggregation algorithms to other datasets, especially those non-binary decision attributes and with more decision makers. We also intend to study the behavior of the aggregation algorithm with large datasets. Finally, we will intend to design new measures to evaluate information loss, precision loss and information redundancy.

References

1. Bi, W.J., Chen, X.H.: An extended dominance-based rough set approach to group decision. In: Guizani, M., Chen, H.H., Zhang, X. (eds.) Proceedings of the International Conference on Wireless Communications, Networking and Mobile Computing (WiCom 2007), Shanghai, China, pp. 5753–5756, 21–25 September 2007
2. Blaszczynski, J., Greco, S., Slowinski, R.: Multi-criteria classification - a new scheme for application of dominance-based decision rules. Eur. J. Oper. Res. **181**(3), 1030–1044 (2007)
3. Chakhar, S., Ishizaka, A., Labib, A., Saad, I.: Dominance-based rough set approach for group decisions. Eur. J. Oper. Res. **251**(1), 206–224 (2016)
4. Chakhar, S., Saad, I.: Dominance-based rough set approach for groups in multicriteria classification. Decis. Support Syst. **54**(1), 372–380 (2012)

5. Chen, Y., Hipel, K., Kilgour, D.: A decision rule aggregation approach to multiple criteria group decision support. In: Proceedings of the IEEE International Conference on Systems, Man and Cybernetics (SMC 2008), pp. 2514–2518. Institute of Electrical and Electronics Engineers (IEEE), Singapore, 12–15 October 2008

6. Chen, Y., Kilgour, D., Hipel, K.: A decision rule aggregation approach to multiple criteria-multiple participant sorting. Group Decis. Negot. **21**, 727–745 (2012)

7. Greco, S., Matarazzo, B., Slowinski, R.: Rough sets theory for multicriteria decision analysis. Eur. J. Oper. Res. **129**(1), 1–47 (2001)

8. Greco, S., Matarazzo, B., Słowiński, R.: Dominance-based rough set approach to decision involving multiple decision makers. In: Greco, S., et al. (eds.) RSCTC 2006. LNCS (LNAI), vol. 4259, pp. 306–317. Springer, Heidelberg (2006). https://doi.org/10.1007/11908029_33

9. Greco, S., Matarazzo, B., Slowinski, R., Stefanowski, J.: An algorithm for induction of decision rules consistent with the dominance principle. In: Ziarko, W., Yao, Y. (eds.) RSCTC 2000. LNCS (LNAI), vol. 2005, pp. 304–313. Springer, Heidelberg (2001). https://doi.org/10.1007/3-540-45554-X_37

10. Pawlak, Z.: Rough Set. Theoretical Aspects of Reasoning About Data. Kluwer Academic Publishers, Dordrecht (1990)

11. Saad, I., Chakhar, S.: Multicriteria methodology based on majority principle for collective identification of company's valuable knowledge. Knowl. Manag. Res. Practice **10**(4), 380–391 (2012)

12. Xu, Z.: Multiple-attribute group decision making with different formats of preference information on attributes. IEEE Trans. Syst. Man Cybern. Part B: Cybern. **37**(6), 1500–1511 (2007)

13. Zhang, H., Dong, Y., Chiclana, F., Yu, S.: Consensus efficiency in group decision making: a comprehensive comparative study and its optimal design. Eur. J. Oper. Res. **275**(2), 580–598 (2019)

Collaborative Decision Making Processes

An Ontology for Collaborative Decision Making

Jacqueline Konaté[1]([✉]) [iD], Pascale Zaraté[2,3] [iD], Aminata Gueye[1] [iD], and Guy Camilleri[2,3] [iD]

[1] Faculté des Sciences et Techniques, Université des Sciences, des Techniques et des Technologies de Bamako, Colline de Badalabougou, Bamako, Mali
`jacqueline.konate@usttb.edu.ml`,
`aminata.gueye@fst-usttb-edu.ml`
[2] IRIT, 118 route de Narbonne, 31062 Toulouse Cedex 9, France
`{zarate,guy.camilleri}@irit.fr`
[3] Université de Toulouse; UPS, INSA, INP, ISAE; LAAS, 31077 Toulouse, France

Abstract. This article focuses on an ontology construction for collaborative decision making. To do this, a state of the art on collaborative decision-making, on ontology engineering and on collaboration engineering has been done. An eight-step ontology development methodology was adopted and implemented to build the ontology. A corpus made up of more than seventy-seven (77) documents was the starting point for the extraction of terms from the ontology and the UML (Unified Modeling Language) language served as a description language of our ontology. This ontology is intended to be the starting point for a facilitation support system in a Collaborative Decision Making process. The aim of the work is to produce a new system according the "Facilitator in the box" paradigm.

Keywords: Ontology · Collaborative Decision Making · Artificial Intelligence · Facilitation · Collaboration Engineering

1 Introduction

Among the collaborative practices in organizations, decision-making is one of the most common and particularly important. This decision-making is done most often in a collaborative way, hence the Collaborative Decision Making. Making collaborative decision-making can have significant benefits such as efficiency, taking into account all stakeholder proposals. However, a collaborative Decision Support System (GDSS) and the attitudes that encourage interaction and sharing are necessary to achieve these benefits. To do this, a number of approaches have been developed; most of them emphasize the need to use an automated tool. However, the use of predefined processes is essential because it allows gaining a substantial advantage of all the resources (human, technology, time, …) mobilized for the decision making. An approach called Collaboration Engineering provides tools for designing collaborative processes for

D. C. Morais et al. (Eds.): GDN 2020, LNBIP 388, pp. 179–191, 2020.
https://doi.org/10.1007/978-3-030-48641-9_13

practitioners that can do without the intervention of a facilitator when using these processes [6].

Facilitation activities are central to decision-making. However, skilled facilitators are rare in organizations as they tend to become managers very quickly [6]. Our ambition is to embed facilitation skills into a system. Several works have already been made to support the activities of the participants in the decision-making process. However, efforts still need to be made to support the facilitation tasks including the parameterization of the tool. This paradigm is called "Facilitator in the box" and intends to encapsulate the skills of a facilitator in a system (GSS: Group Support System) that can self-configure according to a number of parameters [4]. In the area of decision-making, this system is called GDSS.

Designing a system that can withstand key facilitation tasks and self-configure requires a lot of conceptualization and prior definition of inference rules. Such a system is an artificial intelligence system and we considered that the development of ontology is relevant. This ontology has to be answered the necessary questions for an automatic parameterization on the basis of certain entries.

Ontology defines a common vocabulary for researchers who need to share information in a field [17].

Ontologies are used in several fields including Philosophy, Linguistics and Artificial Intelligence (AI). As the goal of AI is to make machines sophisticated enough to integrate the sense of information [16], the organization of knowledge is a very important step towards achieving this goal. This step gave birth to knowledge engineering that relies heavily on ontologies as a means of representation and organization of knowledge. The goal of our work is to conceptualize the field of collaborative decision-making to develop a collaborative decision-support tool. Thus, the definitions of ontology to which we will be concerned are related to the discipline of Artificial Intelligence.

The ontology definition language we used is UML (Unified Modeling Language) which is initially designed for the representation of the structure of a system. However, it has been shown that UML has many similarities with OWL (Web Ontology Language) and that the first one can be used in place of the second one. Cranefield is the precursor of this approach [9].

In this context, this article presents a work whose objectives are:

- the characterization of the different types of ontologies,
- the identification of the type of ontology that fits to our needs,
- the selection of a methodology for the construction of an ontology,
- the implementation of this methodology for an ontology development for collaborative decision-making.

The remain of the paper is organized as follows: Sect. 2 presents the state of the art in collaborative decision-making, ontological engineering and collaboration engineering; Sect. 3 presents the ontology implemented through the methodology and finally Sect. 4 presents the conclusions of this work and the associated perspectives.

2 Background

2.1 Collaborative Decision Making Process

In this section, we are going to highlight the main phases of a generic collaborative decision making process. Figure 1 presents a general view of it.

The **preparation phase** consists of defining the problem of decision-making according to the ontology. In other words, we must define the purpose, the domain, the current context of the problem as well as the criteria and possible constraints.

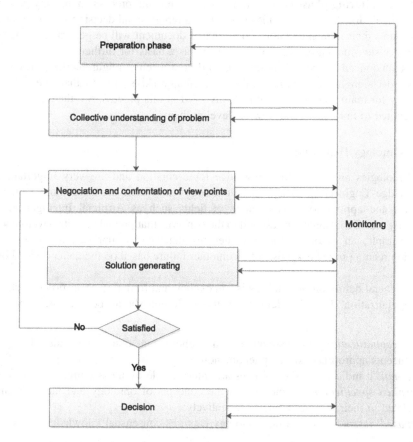

Fig. 1. Generic model of the collective decision-making process [1].

The **phase of collective understanding of problem** can be considered as an extension of the preparation phase. Indeed, it consists of sharing a common vision of the problem with all the participants and finding an agreement on how to implement the designed process [19].

The **solution generation phase** is the beginning of the treatment and it consists to produce alternative ideas to solve the problem.

The **phase of negotiation and confrontation of viewpoints** comes after that of generation to allow participants to elaborate their contributions by arguing them in order to win the support of the greatest number.

The **decision phase** occurs after the negotiation and confrontation phase of the solutions. It consists in selecting, according to the criteria previously defined, the ideas which have been approved by maximum participants, or which will have made the consensus within the group.

The **monitoring phase** covers all the decision making process so that any problem can be timely fixed. It also includes generating a report on all decision-making process and ensures their implementation. To do that, a document will be generated at the end of the decision-making process and will serve as a basis for further work.

As announced below, our purpose is to design a self-configurable tool for collaborative decision making which embedded faciltator skills. To do that, we need an ontology for main concepts of domains involved in this process. So, the next sections are devoted to ontologies and how to develop them.

2.2   Ontology Definition

The ontologies are useful for knowledge representation and are very important for Knowledge Engineering. The name «ontology» has been used for more than two decades and applies to a wide range of fields such as Artificial Intelligence. An ontology must be designed to respond to the concerns that prevailed in its development. For example, an ontology for computer maintenance is supposed to answer any question relating to the diagnosis of a computer failure based on the data provided by a user.

A general definition of ontology is "an explicit and formal specification of a shared conceptualization" [3]. The elements of this definition should be understood as such [16]:

- *conceptualization*: an abstraction of a phenomenon, obtained by identifying the concepts appropriate to this phenomenon;
- *formal*: it indicates that ontologies are interpretable by the machine;
- *explicit specification*: it means that the concepts of ontology and the constraints related to their use are defined declaratively;
- *shared*: refers to the fact that ontology captures consensual knowledge.

Through the elements above, we can deduce an ontological terminology, a common vocabulary, a domain name, the concepts and the relations between them.

In the next subsection, we present the different types of existing ontologies and their characteristics.

2.3   Types of Ontologies

In this section we do not intend to give an in-depth typology of ontologies. On the other hand, we want to give a general view on the main types of ontologies [13]:

- **High level ontology:** this type of ontology presents concepts regardless of context or situation. In doing so, since the concepts are out of context, they must be widely accepted in general by people. All or almost everything could be represented using this approach.
- **Domain Ontology:** as its name suggests, this type of ontology is conceived in a specific context. It is an organization of knowledge related to an area that could be agriculture, livestock, civil engineering.
- **Tasks Ontology:** this type of ontology is used in problem-solving context to represent concepts related to the specific tasks that are executed. These tasks may be those related to the design, those related to a surgical operation.
- **Application ontology:** an application ontology is quite particular because it associates two specific ontologies, one of which is a domain ontology and another one is a task ontology. In other words, it is a conceptualization of the tasks carried out by the actors of the field when they are at work.

Since our goal is only to design an ontology for a particular domain: the collaborative decision-making. Our ontology will be thus of the fourth type: an application ontology with a domain straddling decision-making and Collaboration Engineering.

In the next section, we will discuss the process of building an ontology.

2.4   Ontology Engineering

To conceive ontology, it is necessary to determine the reasons to build ontology and to adopt a methodology to construct this ontology. In the following, we present these two steps.

Reasons to develop ontology. According to Noy and McGuinness [17], the needs to develop an ontology are diverse and varied and we can mention:

- *The need for a common understanding among software designers*
  According to Gruber, this reason is one of the most common reasons to develop ontologies [12]. Indeed, although software is designed for a given domain, its use is transversal to several domains. For example, there are websites that manage online bookstores. These web sites can share the same ontology, then search engines can easily extract information from these different sites.
- *The need for reuse of knowledge on a domain*
  This need has been instrumental in advancing ontology research. Indeed, cross-domain notions tend to be defined differently across domains. When specialists agree to design general ontology for users that can be used as basis for specific ontologies.
- *Make explicit what is considered implicit on a domain*
  There are several notions that are perceived differently by the actors of a domain because they are not clearly defined. To help stakeholders better understand themselves or even to increase their knowledge of the field, it is important to explicit the notions.

- *The need to distinguish knowledge on a field and operational knowledge*
  This is the difference between a domain ontology and a task ontology of the same domain. The first one is a representation of the concepts, so the knowledge on the domain and the second one is a representation of the operational concepts of the domain.
- *Analyze knowledge on a domain*
  In essence, ontology gives details on concepts, especially the important concepts are explicitly and accurately defined. It is naturally a propitious basis for the formal analysis of the concepts of the field in which the ontology has been elaborated.
  It is important to see that domain ontology is not always a goal. However, the concepts that it contains and their structure are likely to be used by other programs [17].
  The following section presents the methodology for building an ontology.

A Methodology for Ontology Building. Works done by Hernandez and Mothe [14] and Lechchine [16] allow to distinguish eight (8) stages for ontology building:

- **Step 1: Ontology requirements specification:** this step consists in determining the domain of knowledge of the ontology, its type, the objectives which it aims and its future use. It also determines the scope and techniques that must be used for concept extraction.
- **Step 2: Choosing the corpus:** this is to select documents from which the concepts will be extracted.
- **Stage 3: Linguistic study of the corpus to extract the terms and their relations:** this step allow to identify the terms representative of the domain as well as the relations which bind them.
- **Step 4: Normalization of the results of step 3:** it consists in identifying and defining the concepts and the semantic relations between the concepts and the terms.
- **Step 5: Modeling the ontology:** it consists of using an ontology language to formally represent the semantic network deployed in step 4.
- **Step 6: Ontology structure building:** it is drawing up a hierarchy of ontology concepts without redundancies and ambiguity and possible new relations.
- **Step 7: Validation of Ontology by domain expert(s):** the work from the previous stage must be approved by the specialists in the field.
- **Step 8: Ontology updating:** Since any domain is subject to change/ evolution, it is important that the related ontologies are regularly updated either by adding new concepts or by reformulating others.

2.5   Collaboration Engineering

Collaboration Engineering is an approach to design collaborative work practices for high value recurrent tasks, and the deployment of these designs for practitioners to perform them themselves without the need for collaboration professional facilitator [6]. This approach is very interesting for us because its objective is to empower participants in a collaborative decision-making process. So, the concepts it promotes can be included in the design of the decision-making tool that we have to design.

Collaboration Engineering aims to provide some of the benefits of professional facilitation to groups that do not have access to professional facilitators.

Some Key Concepts. Here, we describe the concepts of Collaboration Engineering approach [4, 7, 10, 15].

- *Collaboration* is joint effort toward a common goal. It is necessary that the participants in the collaboration make the effort to achieve the goal of the group.
- *Goal* is a desired state or result.
- *Work practice* is a set of actions performed repetitively to accomplish a particular organizational task.
- *Collaborative task* is one that success depends on the joint efforts of several individuals.
- *High value task* is one that brings substantial advantage to an organization or one that avoids a substantial loss by its successful completion. The practice of the task produces a high value over time for organization.
- *Recurrent task* is one that can be conducted repeatedly through a similar process each time it is executed. In other words, it indicates that a similar approach can be used each time the task is executed even with some variations in its parameters.
- *Collaboration Process* is a series of activities performed by a team for a specific purpose within a given time frame.
- *Facilitator* is someone who both designs and conducts a dynamic process that involves managing relationships, tasks and technology, as well as structuring tasks and contributing to the achievement of the outcome of collaboration process. This work is called Facilitation.
- *Practitioner* is a specialist in a task and must perform important collaborative tasks such as risk assessment or decision making as part of his or her professional duties. He executes a specific collaborative process on a recurring basis, so he does not need any facilitation skill.
- *Collaboration Engineer* is the one who designs and documents collaboration processes that can be easily transferred to a practitioner. This means that the practitioner can run a process without the help of a collaboration engineer or facilitator.
- *ThinkLet* is the smallest unit of intellectual capital needed to create a collaboration pattern.
- *Reusability* is the property of a thinkLet to be used to solve problems different from those for which they were originally created.
- *Predictability* is the property of a thinkLet, when executed as prescribed, to create similar variations in overall collaboration patterns and deliverables with a variety of teams, tasks, and conditions.
- *Transferability* expresses the degree to which people who have never created a thinkLet can learn it, remember it, and execute it successfully.

About thinkLets. Design patterns were introduced by Alexander who is architect by a 1970s observed a recurrence of problems that occur in the architectural design phase. He imagined the concept of pattern as follows: "A pattern describes a problem that repeats and repeats itself again and therefore describes the core of the solution to this problem, so that you can use this solution million times without never do it the same way twice" [2]. Existing patterns are about two hundred and fifty three (253) and cover all aspects of building construction.

Pattern concept was adopted for software design by Gamma [11] who helped to prove the value of the concept and made him a reference in computing field.

The concept of design pattern has also been taken up by collaboration engineering theorists to propose a new Design Pattern language for collaboration called thinkLets. They are the best practices of expert facilitators to support groups in their collaborative efforts to achieve goals. ThinkLet is defined as the smallest unit of intellectual capital needed to create a collaborative pattern [6]. They are reusable, predictable and transferable facilitation techniques that can be used to drive a group through a process towards its goal [10]. Each thinkLet is an instantiation of one following six general patterns, also with sub patterns [7]:

- "Generate" allows the move from having few to having more concepts,
- "Clarify" allows moving from having less to having more shared meaning of the concepts under consideration,
- "Reduce" is the opposite of the Generate pattern,
- "Organize" is moving from having less to having more understanding of the relationships among the concepts,
- "Evaluate" is to move from having less to having more understanding of the utility of the concepts priority with respect to goal attainment.,
- "Build Consensus" is to move from having more to having less disagreement on the course of action.

For more details on thinkLets, see the paper from Briggs and de Vreede [5].

In the next section, we present how to design a collaboration process based on the concepts we presented earlier.

3   Developing an Ontology for Collaborative Decision Making

Our goal is to design a GDSS with the possibility of setting it automatically. In this part, we will present the different steps that we followed to develop our ontology, and then we will present our ontology.

3.1   Our Ontology Building

In this part, we will identify the questions that our ontology must answer, then apply the methodology presented in Sect. 2.4 to build our ontology.

Questions Must be Answered by Ontology. After a process setting effort, five (5) main parameters were identified. These are: (1) goal of decision, (2) purpose and scope of decision making, (3) number of decision makers, (4) process duration, and (5) anonymity. Based on these parameters, the questions to which the ontology must respond have been developed. They are presented below in order of priority:

1. Is decision-making multi-criteria or single-criteria?
2. Which collaboration patterns of should be used for this process?
3. What kinds of decision makers (skills, qualities, characteristics, …) should participate in this process?
4. What is the appropriate number of decision makers?
5. What should be the duration of the process?
6. Is anonymity required for this process?

Above questions bring to understand that the reason to develop the ontology is «*The need to distinguish knowledge on a field and operational knowledge*» according to Sect. 2.4 presented above. We will proceed to apply the methodology to build an ontology that meets the needs specified above.

Methodology Implementing.

Step 1: Ontology Specification

Our ontology is between several **areas of knowledge** that are: Decision Making (DM), Decision Support Systems (GSS, GDSS) and Collaboration Engineering (CE).

The type of ontology is **Application Ontology:** two domain ontologies have been selected and are: the GDSS and CI domains and one tasks ontology which is the domain of Decision Making (DM)

The **objectives** for the ontology are essentially: identify the different parameters of a decision process and identify the appropriate collaboration patterns (thinkLets) for a given situation

The **scope of the ontology** covers the important concepts of Decision Making, Decision Support Tools (GDSS) and Collaboration Engineering (CE).

The **techniques** used to develop ontology are extracting concepts, relationships and axioms.

About the **use of the ontology**, it concerns the facilitation, the automatic parameterization of the collaborative decision-making assistance tools (GDSS).

Step 2: Choosing the corpus:

We have selected more than 177 papers consisting of research papers and technical reports from various sources including conference proceedings and journals through Google Scholar, IEEE Xplorer, ACM, JSTOR, ScienceDirect, ResearchGate, GDN, and also direct contact with authors. We used keywords like: Collaboration, Decision Making, Collaboration Engineering, Groupwares, DSS and GDSS.

Step 3: Linguistic study of the corpus to extract representative terms of the domain and their relations (lexical and syntactical)

We used the extraction tool named Termostat to extract the significant terms of the different domains as well as the relationship between them.

Step 4: Normalize the results of step 3

After extracting the concepts from the previous step, we identified those that contributed to answering the six (6) questions identified in Subsect. 3.1.1. Then, we defined these concepts and related them each other.

Steps 5 & 6: Ontology Modeling and Building Ontology Structure

To formally represent ontology, we used UML (Unified Modeling Language) to represent the concepts and relationships from the previous step and to build the hierarchy of concepts for the three domains. Indeed, there are several works in the literature that argue UML can be used as an ontology description language because of their similarity [8, 9, 18]. In addition, we selected this solution because it allowed us to merge steps 5 and 6 of the methodology. In addition, UML would facilitate the implementation of the group decision making tool.

Ontology does not contain redundancies in relationships. Each concept has been defined explicitly to remove any ambiguity in its meaning. Similarly, the relationships between concepts are named to help understanding and the enrich ontology. For more details on this work, see Sect. 3.2.

Step 7: Validation of Ontology by domain expert(s)

Ontology was evaluated by specialists from domains of knowledge concerned with ontology: Decision Making, Collaboration Engineering, Group Decision Support Systems and Ontology Engineering. Each of the experts made comments and observations on the work done. Efforts have been made to improve work according to their contributions (see section 3.2).

Step 8: Update Ontology

For the evolution of our ontology, we plan to set up a community around the proposal. This community will animate the various activities aimed at adapting the ontology to the needs of each other.

3.2   An Ontology for Collaborative Decision Making

The implementation of the methodology produced a result whose schematic representation is given in Fig. 2.

1. **Decision Problem:** the subject for which a decision must be made. For example, there may be resistance from users to adopt a new system.
2. **Constraint:** A requirement that must be satisfied by the process or an element of the process. For example, the minimum or maximum number of participants in the decision-making process.
3. **Criterion:** an element allowing making a decision.
4. **Decision making process and subProcess:** this term refers to the successive states through which the group passes to arrive at the decision. It is a continuation of different phases of a decision-making.
5. **Report:** provides a detailed presentation, a restitution of all activities, and also contains the final result (the adopted decision).
6. **Alternative:** a choice, a possible solution, a probabilistic result provided by the decision makers. Choices having different consequences, the final choice will be made according to specified criteria.
7. **Decision_Result:** it is the final choice, the adopted alternatives, the (optimal) options selected according to the criteria (if there is any).
8. **Resource:** Technology, tools and other physical resources such as money or a meeting room can be used in this process.
9. **Phase and Reunion:** is a step in a process that may consist of one or more subphases that will be called Reunion. A reunion represents the various groups, the assemblies of people created to discuss about topics of decision-making.

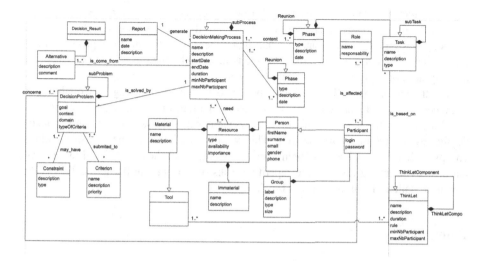

Fig. 2. An ontology for collaborative decision making

10. **Task and subTask:** is a job to be done by a participant. A task can have of subtasks.
11. **Role:** the function, the right of use that a user has, the different tasks that he must perform. One of the important roles is that of the facilitator who is a participant that drives the decision-making meetings.
12. **Participant** (or agent or actor or stakeholder or user): is any individual who intervenes and plays a role in decision-making.
13. **Group:** is a set of decision-makers with common factors, a collective goal that collaborates in decision-making.
14. **Person** (or human or human resource): characterized by first name, surname, gender, email, phone number.
15. **Material:** nonhuman material resource.
16. **Immaterial:** non-material and non-human resource.
17. **Tool:** means used to perform one or more parts of the decision-making process.
18. **ThinkLet, ThinkLet_Component, ThinkLet_Composite:** refers to the different collaboration patterns used in decision-making. The ThinkLet can be compound or elemental.

4   Conclusion

This article presents ontology for collaborative decision-making. It is an application ontology built from the following areas: collaborative decision-making, decision support systems, collaboration engineering. The development approach that we adopted has eight stages and the starting corpus consists of seventy-seven (77) journal articles, conferences papers and technical reports from Google Scholar, Elsevier, IEEE explorer,

ScienceDirect, etc. The tools used are Termostat for terms extraction from the corpus and UML for ontology hierarchy building and its standardization. The resulting ontology has eighteen (18) concepts that have been defined with their properties and thier relationships. The use of this language is justified by the purpose of our work: the development of a group decision support system.

For the validation of the ontology, specialists in decision-making, in Collaboration Engineering, in collaborative tools for decision-making, in ontology engineering.

Our goal is to build a collaborative decision support system that automatically configures itself based some entries. This would greatly assist the decision-making groups in their work by giving them some autonomy from the professional facilitators, but also by relieving them of the repetitive tasks of configuring the tools. This is the paradigm called "Facilitator in the box".

References

1. Adla K.: Aide à la Facilitation pour une prise de décision Collective: Proposition d'un Modèle et d'un Outil. Thèse de l'Université de Toulouse, soutenue le, 8 June 2010 (2010)
2. Alexander, C., Ishikawa, S., Silverstein, M.: A Pattern Language: Towns, Buildings, Construction. Oxford University Press, New York (1977)
3. Borst, W.: Construction of Engineering Ontologies. Unpublished doctoral dissertation, University of Tweenty (1997)
4. Briggs, R.O., Kolfschoten, G.L., De Vreede, G.J., Lukosch, S., Albrecht, C.C.: Facilitator-in-a-box: process support applications to help practitioners realize the potential of collaboration technology. J. Manag. Inf. Syst. 29(4), 159–194 (2013)
5. Briggs, R., de Vreede, G.-J.: ThinkLets: Building Blocks for Concerted Collaboration. Briggs and de Vreede, Nebraska (2009)
6. Briggs, R.O., De Vreede, G.J., Nunamaker, J.F.: Collaboration engineering with thinklets to pursue sustained success with group support systems. J. Manag. Inf. Syst. 19(4), 31–64 (2003)
7. Briggs, R.O., Kolfschoten, G.L., de Vreede, G.-J., Dean, D.L.: Defining key concepts for collaboration engineering. In: Rodríguez-Abitia, G.I.A.B. (ed.) AMCIS, vol. 17 (2006)
8. Brockmans, S., Volz, R., Eberhart, A., Löffler, P.: Visual Modeling of OWL DL ontologies using UML. In: International Semantic Web Conference, pp. 198–213 (2004)
9. Cranefield, S.: Networked knowledge representation and exchange using UML and RDF. J. Digit. Inf. 1 (2001)
10. De Vreede, G.J., Briggs, R.O.: Collaboration engineering: designing repeatable processes for high-value collaborative tasks. In: Proceedings of the Annual Hawaii International Conference on System Sciences (2005)
11. Gamma, E., Helm, R., Johnson, R., Vlissides, J.: Design Pattern. Addison-Wesley, Boston (1995)
12. Gruber, T.R.: A translation approach to portable ontology specifications. Knowl. Acquis. 6, 199–221 (1993)
13. Guarino, N.: Formal ontology and information systems. In: Proceedings of Formal Ontology in Information System, pp. 3–15. IOS Press (1998)
14. Hernandez, N., Mothe, J.: TtoO: une méthodologie de construction d'ontologie de domaine à partir d'un thésaurus et d'un corpus de référence, Rapport interne, IRIT (2006)

15. Kolfschoten, G.L., Briggs, R.O., de Vreede, G.J., Jacobs, P.H.M., Appelman, J.H.: A conceptual foundation of the thinkLet concept for Collaboration Engineering. Int. J. Hum Comput Stud. **64**(7), 611–621 (2006)
16. Lekhchine, R.: Construction d'une ontologie pour le domaine de la sécurité: Application aux agents mobiles, Rapport de master, Université de Mentouri, Algerie (2009)
17. Noy, N.F., McGuinness, D.L.: Ontology development 101: a guide to creating your first ontology. stanford knowledge systems laboratory technical report KSL-01-05 and stanford medical informatics technical report SMI-2001-0880, March 2001 (2001)
18. Pătrașcu A.: Comparative analysis between OWL modelling and UML modelling. Econ. Insights Trends Chall. **IV(LXVII)**(2), 87–94 (2015)
19. de Terssac, G., Chabaud, C.: Référentiel opératif commun et fiabilité. In: Leplat, J., de Terssac, G., (eds.) Les facteurs humains de la fiabilité dans les systèmes complexes, pp. 111–139. Octarès, Toulouse (1990)

Decidio: A Pilot Implementation and User Study of a Novel Decision-Support System

Kevin Hogan[✉], Fei Shan, Monikka Ravichandran, Aadesh Bagmar,
James Wang, Adam Sarsony, and James Purtilo

Department of Computer Science, University of Maryland,
College Park, MD 20742, USA
khogan@cs.umd.edu

Abstract. In this work, we add to the rich history of decision-support system research by implementing and evaluating a pilot implementation of a novel system, which we call Decidio. Our tool was integrated into a pre-existing decision-making process regularly conducted by 9 teams of undergraduate students. We find an overall positive response to Decidio based on the results of a tool evaluation survey that we conducted after our experiment. Furthermore, we conduct a Big-Five Factor personality survey of participants and associate personalities with interactions recorded by our tool. We find that the students who demonstrate leadership behaviors through their interactions score higher in extraversion and lower in conscientiousness than other students. Our analysis also reveals that agreeableness is strongly correlated with dissimilarity between group ranking outcomes and initially indicated individual preferences.

Keywords: Collaborative decision making · Decision-support system

1 Introduction

Team dynamics and group decision-making are central to any organization. Researchers have reviewed and elucidated what years of research on small groups and teams can tell us about the processes that contribute to team effectiveness. Using that knowledge they have identified leverage points for making teams more effective such as team learning, cohesion, efficacy and group potency [8]. Furthermore, a study examining relationships among work teams with respect to team composition (ability and personality), team process (social cohesion), and

This study was performed after being approved by the Institutional Review Board, University of Maryland, College Park with Ref. no. 1505396-1. This work was supported by the United States Office of Naval Research under Contract N000141812767.

© Springer Nature Switzerland AG 2020
D. C. Morais et al. (Eds.): GDN 2020, LNBIP 388, pp. 192–204, 2020.
https://doi.org/10.1007/978-3-030-48641-9_14

team outcomes (team viability and team performance) found that interpersonally oriented personality characteristics, such as conscientiousness, agreeableness, extraversion, and emotional stability can also be important predictors of team effectiveness [1].

In the 1970s, decision support systems (DSS) were described as "interactive computer-based systems, which help decision makers utilize data and models to solve unstructured problems"[6]. Group decision support systems (GDSS) eventually evolved to consider not only the role of the ultimate decision-maker, but also the communication between everyone in the decision-making group [3,5,14]. Turban et al. [14] have identified the major components that make up these systems. Some of them, such as database analysis and visualization, constitute independent fields of research within computer science. Others, such as support for Multi-Criteria Decision Making (MCDM), more appropriately fall under management science [13]. In recent years, researchers have explored Internet-based algorithms and workflows for decision-making at scale [9,11]. Furthermore, inferring consensus among members of a group in a data-driven fashion has long been a goal in anthropology and social psychology. [12] demonstrates the use of Cultural Mixture Modeling (CMM) for inferring consensus among members of heterogeneous multidisciplinary professional teams in an example decision-making process. The findings of their study inform team decision-making based on shared beliefs and team composition.

In this work, we present a pilot DSS of our own design and construction, which we call Decidio. While we plan to extend Decidio to be a general-purpose DSS, our present version of the software is only able to support a very basic collaborative ranking process. Our pilot implementation was designed to support a real collaborative ranking carried out by University of Maryland undergraduate students. Our goals in this work are two-fold. First, we wanted conduct a user study of Decidio to identify its effectiveness and potential areas of improvement. Second, we wanted to use the opportunity to explore group dynamics in the particular decision process that we were supporting.

The remainder of the paper proceeds as follows. In Sect. 2, we describe the functionality of our current implementation of Decidio. In Sect. 3 we present the use case for the user study of Decidio and demonstrate how our tool was incorporated into an existing group decision-making process. In Sects. 4 and 5, we document and discuss the results of our experiment. We measure the reception of the tool by its users through a user survey, and we find the feedback to be positive overall, with some areas of improvement identified. We also use log events from the tool to create behavioral profiles of the users involved in decision-making. Finally, we explore the connection between these behavioral profiles and user personalities.

2   Functionality of the Decidio Pilot Implementation

Decidio is a web-based collaborative ranking to support groups that need to collectively rank a set of options. The tool uses the Python-based Flask web server for its backend and the React Javascript Library for its frontend. This iteration of Decidio was designed to 1) make the ranking of options more efficient, and 2) ensure that the deliberation on team preference would fairly represent the opinions of all members on the team. The tool therefore supports the following features:

1. **Linking users and teams**: Decidio supports grouping users into teams. Every user is assigned to a team.
2. **Adding a list of options**: An administrator can add a list of options which will be visible to all the users.
3. **Individual ranking**: Users can individually view the list of options added by the administrator and can submit their own preferred ranking order. Individual rankings are submitted for every user.
4. **Group ranking**: Groups can collectively choose a final set of rankings in Decidio's group ranking page. Updates made to the ranking by any member in a group are displayed in real-time to everyone in that group.
5. **Displaying statistics about individual rankings**: Decidio displays basic statistics for the group, generated from the individual ranking phase when the users are working on the group rankings. This feature helps users make decisions which encompass opinions of everyone in the group.
6. **Analysis of user behavior through logs**: Decidio logs highly granular interactions in the tool. Based on the logged activity, we can classify users into different categories like operators, leaders, followers, etc. We describe these terms in Sect. 3 in more detail.

3   User Study

The Quality Enhancement Systems and Teams (QUEST) Honors Program is a University of Maryland program for undergraduate students. The first course in the program involves a final project in which teams of students collaborate with a campus organization on process improvement. Each team is able to indicate their preference for which of the available projects they would like to work on. Historically, teams had indicated this preference by submitting a hand-written ranking of projects to the course instructor after a brief period of deliberation. The instructor would then allocate the projects to fulfill these preferences to the best of her ability.

Our team identified this decision-making process as an opportunity to test the pilot of Decidio in a real-world scenario. The tool was used by 45 users divided into 9 teams. There were 9 project options which the teams had to collectively rank in order of preference.

3.1   QUEST Workflow

There are three main phases in the workflow that we designed for QUEST. We describe each phase in detail.

3.1.1   Individual Project Ranking

QUEST students attended a special class session in which the available projects were presented by the instructor. Before arriving to this session, students were required to indicate their individual preferences for projects. The instructor for the course distributed a document describing each of the projects in detail during the days leading up to the session. Students used the Decidio web interface to order these projects from most-desirable to least-desirable. Figure 1 shows the interface and describes its interactivity.

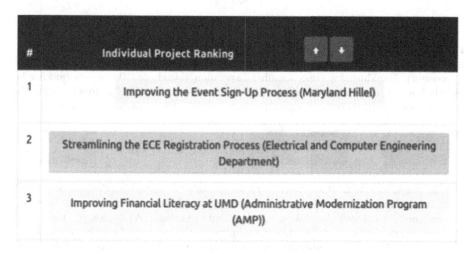

Fig. 1. Dashboard for individual project ranking. Users click on a project to select and then use the up and down arrows to re-order that project. Once they are satisfied with a ranking, they click a submission button to finalize (button not shown in the screenshot).

3.1.2   Group Project Ranking

After the instructor presented each project available for selection, students had 30 min to meet with their groups and decide on a single ranking of projects to submit to the instructor. Each team sat at a separate circular table in the same classroom during the discussion. Students logged into Decidio via their personal laptops and navigated to the group project ranking dashboard. The important components of the dashboard are shown in Fig. 2 and Fig. 3. This dashboard is distinguished from the individual ranking dashboard in that 1) all users in a team have equal control over modifying and submitting a unified project ranking and 2) users are presented with the results of individual rankings for all team members. Below, we describe the flow of group ranking:

1. **Collaborative Ranking.** Each user is able to update the team ranking with the same interaction used for the individual ranking process. When the ranking changes, this change is published to the dashboard of each team member, and the dashboard will automatically refresh to reflect the changes.

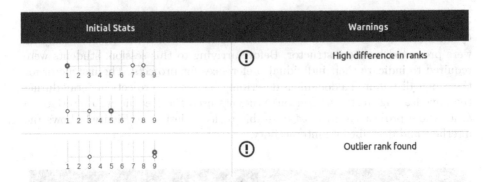

Initial Stats		Warnings
	⚠	High difference in ranks
	⚠	Outlier rank found

Fig. 2. Section of the dashboard for group project ranking that shows the individual rankings for each project (rows are aligned with project names not shown in the screenshot). Rankings for team members are anonymized, but the user's ranking is hightlighted in red as a reminder of their submitted preferences. The tool calls attention to cases where there are large differences in project preferences or outlier ranks.

Fig. 3. Staging controls at the bottom of the group project ranking dashboard. In the configuration shown, the ranking has been staged (the leftmost button toggles between "Stage" and "Unstage" depending on the staging status). While staged, the ranking component shown on the dashboard (similar to the individual ranking component shown in Fig. 1) will be read-only. The user has approved the ranking, and no other member of the team has approved. Since the number of approvals falls short of the team size, Decidio displays a warning icon.

2. **Staging.** Users can choose to stage their team's ranking at any point during Collaborative Ranking by clicking a button on the web page. When a ranking is staged, further modifications to this ranking are frozen and the staged ranking is displayed for review by team members. Any user is able to unstage the ranking after it has been staged, which will return all users to Collaborative Ranking.
3. **Approval and Submission.** When a ranking has been staged, users can click to indicate whether they approve of the ranking for submission. The tool indicates to all users the number of team members who have approved the ranking, warning users when approval is not unanimous. However, the tool does not prevent submission based on the approval status. A user can submit at any time by clicking a button on the dashboard.

3.1.3 Project Assignment

Projects were assigned to teams outside of Decidio. The rankings for each team were presented to the instructor on an administrator dashboard, which allowed for convenient export to Excel. The instructor was then able to apply a technique used in previous semesters for the fair allocation of projects.

3.2 User Activity Logs

Every server request in Decidio is recorded in a log file. A log normally contains the timestamp, user ID, group ID, and the action taken. Actions include information like:

- Has a user submitted their individual ranking?
- What did a user change the group rankings to?
- Which user approved the ranking?
- Which user moved the group ranking to the staging phase?

From the user actions in the log file, we created an activity log that informs us of their interaction patterns with the tool. This profile was later used to classify users into groups based on their actions. Results from the analysis of activity logs are described in Sect. 4.2.

4 Results

4.1 Tool Evaluation Survey

A survey questionnaire (See Appendix) was designed and sent to the students after the group decision making activity. The goal of this survey was to

1. Collect feedback on tool
2. Learn their process, methods, and strategies employed for decision-making when ranking projects as a group
3. Capture some elements of the unstructured discussion that happened in-class for group project ranking

We received 38 responses for the survey which were analyzed, and the findings from the survey are listed below.

1. **Visualizations of individual rankings promoted group discussions**: More generally, the visualizations showing the rankings of group members appeared well accepted by the students. We learned that the visualizations helped them understand the group members' project preferences that led to constructive discussions. One user (U11) described that *"The chart was very helpful, because it showed majority perspectives which allowed everyone's perspective to be objectively viewed. In this way, everyone's opinions were incorporated so that one person's opinion couldn't overshadow others."*
2. **Group members played various roles in the decision making process:** From the survey, we learned that during the group decision-making process students took up roles such as note takers and facilitators. Every student team had at least one member acting as the facilitator. Collectively, this showed us the need and the importance of "roles" in group decision making.

3. **Individual and group project rankings changed with understanding of the projects:** In general, project presentations in class prior to group ranking helped students gain knowledge on projects. Group discussion that followed allowed them to communicate their understanding and validate assumptions with their group. This showed us that the presentations and discussions held in class influenced the group's final project rankings. This validated that a mental model which is an internal representation of knowledge changes with individual's understanding of the present state of a system and that the development of a "shared" understanding of the problem within groups is key in group decision making [4].

4. **Groups used various techniques for deciding on the final project rankings:** As a standard discussion protocol was not established for QUEST, groups used a variety of techniques to arrive at the final project ranking. These include making a pros and cons list (8), voting on projects (6), brainstorming ideas, taking-turns or going around the table to discuss implications and levels of interest for each project (10). U39 described their group's process as *"We first started by saying our top 3 and bottom 2. From there we discussed and found the projects the most people had in common."*

4.2   User Activity Logs

An activity log file created for every user that captured their actions as a workflow diagram was used to understand how student groups used Decidio in their decision making process. We observed varied numbers of actions amongst users ranging from just 2 to 73 in their activity logs.

From analyzing workflows in the activity logs of group members, we noticed that the workflows can be grouped into four categories based on who the users were and the actions they performed in the tool.

- **Absentee Workflow:** This workflow was exhibited by users that did not participate in the group ranking activity. It showed only actions related to the submission of individual project rankings.
  Fig. 4a demonstrates this workflow.
- **Regular Workflow:** The regular workflow is the one where users submit individual rankings, approve and/or submit group project rankings. Their flow is mostly linear and does not involve updating the project rankings, staging the project rankings and unstaging the temporary final rankings. Figure 4b demonstrates this user's workflow.
- **Active Operator Workflow:** Workflows containing repeated updates to project rankings indicate that the user is an "active operator" for the group. Each group typically had one or two members actively updating the project order in Decidio. By mapping active operators to their roles from the survey, we found that 11 out of 17 active operators considered themselves to be the facilitators of their groups. Figure 4c demonstrates an active operator workflow.

(a) Workflow for a user who was absent during the group rankings

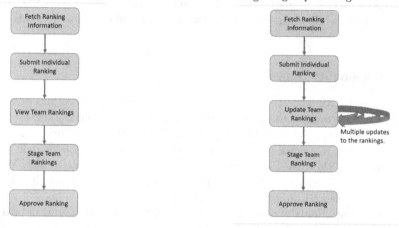

(b) Workflow for a normal user not acting as an operator

(c) Workflow for a user acting as the core operator. The user updates the rankings repeatedly.

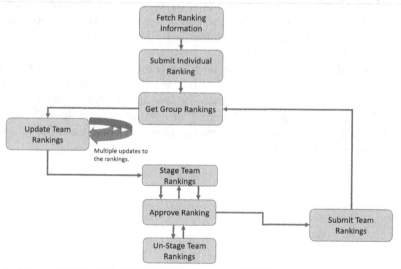

(d) Shows an advanced workflow in a team with the most conflicts. The user is doing multiple actions along with just being the operator in this case.

Fig. 4. Workflows for advanced cases

- **Irregular Workflow:** Irregular workflows contain cycles. i.e, the users have performed staging and unstaging operations. This workflow seems to indicate that the group didn't have the approval of all its group members (no consensus) and had to revise their project rankings by going back and forth between staging and unstaging before proceeding to submit. However, we note that one group exhibiting this workflow claimed via our survey that they had a consistent consensus throughout the ranking process. For this reason, we suspect that there are other explanations for the irregular workflow, such as user errors while navigating the software. Figure 4d demonstrates this type of workflow.

4.3   Personality Survey

Bayram and Aydemir [2] examine the relations of decision styles and personality traits among groups of university students. We use a personality survey and the activity logs from our tool to carry out a similar thread of investigation. We asked students to take a test that measures their "Big Five" personality traits, also known as the Five-Factor Model [10] and collected their results. We then analyzed the 37 responses to study the relations between personality traits of group members and group decision making.

From comparing the personality trait scores of "active operators" from the logs, we learned that the active operators tend to have higher median score on extraversion and lower median score on conscientiousness.(Figure 5b)

We also explored how personality may have affected the decision outcome for each group. We came up with a metric to represent the dissimilarity between two rankings. This is calculated as the sum of squared differences between ranks for each project. For example, the dissimilarity between a ranking of 1-2-3 and 3-2-1 for projects A-B-C would be 8.

The dissimilarity metric cannot fairly be compared between teams without normalization. This is because the distribution of initial rankings varies between teams (some teams naturally had more or less disagreement before the Group Project Ranking phase began). Therefore, we compute a normalized dissimilarity for each user that is the difference between that users raw dissimilarity and the average dissimilarity for that team.

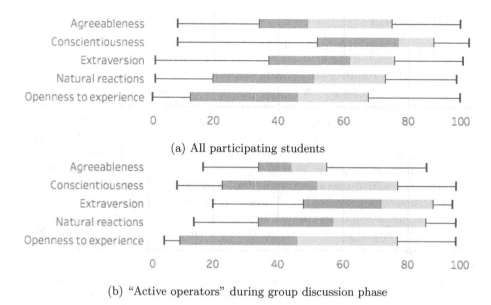

(a) All participating students

(b) "Active operators" during group discussion phase

Fig. 5. Distribution of personality traits of all students and the "active operators" of each group

We looked for correlations between normalized dissimilarity and percentile scores for the Big Five Factor personality survey. We calculated a Pearson correlation coefficient between normalized dissimilarity and each of the five personality factors. The results of this analysis are shown in Table 1. The only strong correlation that we see is the strong positive correlation between Agreeableness and normalized dissimilarity. We interpret this to mean that agreeable students were less likely to see their individual preferences reflected in the rankings submitted for their group.

Table 1. Pearson correlation between Big Five Factor personality characteristics and normalized dissimilarity between individual project ranking and the ultimately selected group ranking. We consider values greater than $+/-$ 0.5 to be strong correlations.

Personality factor	Correlation coefficient
Agreeableness	0.52
Conscientiousness	0.27
Extraversion	-0.01
Natural Reactions	0.03
Openness to experience	0.31

5   Discussion

5.1   Observations Based on User Personalities

The personality comparison between all students and active operators (see Sect. 4.3) is partially aligned with prior research correlating Big Five traits with leadership [7]. Judge et. al. see positive correlations with Extraversion, Openness to experience, and Conscientiousness; and a negative correlation with Natural reactions (referred to as Neuroticism in their work). The bias towards Extraversion among active operators is consistent with these results, however the bias against Conscientiousness in that group is not. Of course, the association we make between active operation of the tool and general leadership is based in assumption and may not hold in all cases (as described in Sect. 5.2).

We also find a strong correlation between agreeableness and our dissimilarity metric. This leads us to the satisfying (and perhaps expected) observation that more agreeable participants were less likely to "get their way" when working towards a team consensus on preference rankings.

We're excited by the insights into decision dynamics that the Decidio pilot provided in this user study, and we plan to enhance Decidio's capability of recording software interactions that are relevant to these dynamics.

5.2   Limitations

We recognize that, because we are documenting results from a single experiment with our software, we should be careful to draw strong conclusions from our work.

There are also several limitations to identify within this one experiment. First, the log events that we collect are not sufficient to capture a rich behavioral profile of all participants in the decision study. We did not record the discussion between each team of participants, and therefore we can not make definitive conclusions about the team dynamic. It's an extrapolation, albeit a reasonable one, to connect the "Operator" workflow described in Sect. 4.2 to the subjective notion of team leadership, for example. We recognize that the "Operator" may have been acting as a sort of team note-taker or secretary for the session rather than the leader of the discussion. Second, we did not test with a control group. This nuances the results from the user survey, as we're not able to draw a comparison to a team who collaborates on a ranking with a default method (e.g., pen and paper). Finally, it is difficult to generalize results about user behavior in the tool when each team consists of a different mix of personalities and individual ranking preferences. For example, a team with insignificant differences between individual preferences for projects may not have much to disagree about, even if their personalities would suggest a higher likelihood of conflict.

5.3   Future Vision

The initial version of Decidio that we present in this paper is only narrowly applicable to group decisions that involve ranking a discrete set of options. However, the ultimate vision for Decidio is to build a DSS that can be easily tailored by users to fit specific decision workflows. This flexibility is valuable to decision-makers because it will allow them to incorporate automation in existing workflows without substantial modification. It is also valuable to researchers who want to experiment with a variety of workflows.

6   Conclusion

In this paper, we present Decidio, a software tool that supports collaborative decision making. We evaluated the tool by using it in real-time with 9 student teams for ranking projects for a class. The results indicate that the overall reception for such a tool is positive. In summary, we learned from our tool evaluation survey that students assume roles and use various techniques when making group decisions. The activity logs showed us the various workflows used by student groups for decision-making. It was interesting to see the relations between personality traits and group decision making. We believe the results from this study will help us in the design and development of a robust decision support system for collaborative decision making.

A   Appendix

A.1   Tool Evaluation Survey

1. What was your first reaction to the tool?
   (A list of 5 reactions with a 1–5 Likert-type scale)
2. Did you or your team face any technical difficulties when using the tool?
   (Yes/no question, followed by a space for open comments)
3. The top two things I like about this interface are...
4. The top two things I dislike about this interface are...
5. In what ways would you modify this tool to increase its usability or functionality
6. How was the chart showing individual rankings of your team members used to stimulate discussion?
7. How were the notes from individual rankings used when ranking projects as a team?
8. What was your role in the team when discussing project rankings?
9. How did the in-class presentations influence your team's project rankings?
10. What aspects of the projects were important to your team when ranking them?
11. How did your team handle disagreements or objections when ranking projects?
12. Would you say everyone in the team had the chance to voice their opinions? If no, what according to you, were the challenges that kept your team from hearing out everyone's opinions?

13. What were the tools or techniques used by your team when making decisions or in resolving conflicts?
14. Which of the following would you say was your team's decision-making strategy
    (A list of strategies such as Consensus, Compromise, Competition, Accommodating, followed by a space for open comments).

References

1. Barrick, M.R., Stewart, G.L., Neubert, M.J., Mount, M.K.: Relating member ability and personality to work-team processes and team effectiveness. J. Appl. Psychol. **83**(3), 377 (1998)
2. Bayram, N., Aydemir, M.: Decision-making styles and personality traits. Int. J. Recent Adv. Organ. Behav. Decis. Sci. **3**, 905–915 (2017)
3. Bui, T.X., Jarke, M.: Communications design for Co-Op: a group decision support system. ACM Trans. Inf. Syst. **4**(2), 81–103 (1986). https://doi.org/10.1145/6168.6169
4. Converse, S., Cannon-Bowers, J., Salas, E.: Shared mental models in expert team decision making. Individ. Group Decis. Making: Curr. Issues **221**, 221–246 (1993)
5. DeSanctis, G., Gallupe, R.B.: A foundation for the study of group decision support systems. Manage. Sci. **33**(5), 589–609 (1987). https://doi.org/10.1287/mnsc.33.5.589
6. Gorry, G.A., Scott Morton, M.S.: A framework for management information systems. Sloan Manag. Rev. **13**(1), 55–70 (1971)
7. Judge, T.A., Bono, J.E., Ilies, R., Gerhardt, M.W.: Personality and leadership: a qualitative and quantitative review. J. Appl. Psychol. **87**(4), 765 (2002)
8. Kozlowski, S.W., Ilgen, D.R.: Enhancing the effectiveness of work groups and teams. Psychol. Sci. Public Interest **7**(3), 77–124 (2006)
9. Leyva López, J.C., Álvarez Carrillo, P.A., Gastélum Chavira, D.A., Solano Noriega, J.J.: A web-based group decision support system for multicriteria ranking problems. Oper. Res. Int. J. **17**(2), 499–534 (2016). https://doi.org/10.1007/s12351-016-0234-0
10. McCrae, R.R., John, O.P.: An introduction to the five-factor model and its applications. J. Pers. **60**(2), 175–215 (1992)
11. Morente-Molinera, J., Kou, G., Pérez, I., Samuylov, K., Selamat, A., Herrera-Viedma, E.: A group decision making support system for the web: how to work in environments with a high number of participants and alternatives. Appl. Soft Comput. **68**, 191–201 (2018). https://doi.org/10.1016/j.asoc.2018.03.047, http://www.sciencedirect.com/science/article/pii/S1568494618301789
12. Perelman, B.S., Dorton, S.L., Harper, S.: Identifying consensus in heterogeneous multidisciplinary professional teams. In: 2019 IEEE Conference on Cognitive and Computational Aspects of Situation Management (CogSIMA), pp. 127–133. IEEE (2019)
13. Triantaphyllou, E.: Multi-Criteria Decision Making Methods: A Comparative Study, vol. 44. Kluwer Academic Publishers, Norwell (2000). https://doi.org/10.1007/978-1-4757-3157-6
14. Turban, E., Sharda, R., Delen, D.: Decision Support and Business Intelligence Systems, 9th edn. Prentice Hall Press, Upper Saddle River (2010)

Author Index

Printed in the United States
By Bookmasters